Cultural Politics and the Transatlantic Divide over GMOs

Cultural Politics and the Transatlantic Divide over GMOs

Hannes R. Stephan
Lecturer in Environmental Politics and Policy, University of Stirling, UK

palgrave
macmillan

First published 2015 by
PALGRAVE MACMILLAN

Palgrave Macmillan in the UK is an imprint of Macmillan Publishers Limited, registered in England, company number 785998, of Houndmills, Basingstoke, Hampshire RG21 6XS.

Palgrave Macmillan in the US is a division of St Martin's Press LLC, 175 Fifth Avenue, New York, NY 10010.

Palgrave Macmillan is the global academic imprint of the above companies and has companies and representatives throughout the world.

Palgrave® and Macmillan® are registered trademarks in the United States, the United Kingdom, Europe and other countries.

ISBN 978–0–230–28457–9

This book is printed on paper suitable for recycling and made from fully managed and sustained forest sources. Logging, pulping and manufacturing processes are expected to conform to the environmental regulations of the country of origin.

A catalogue record for this book is available from the British Library.

Library of Congress Cataloging-in-Publication Data
Stephan, Hannes R., 1978– author.
Cultural politics and the GMO divide : cultures of nature / Hannes R. Stephan, lecturer in Environmental Politics and Policy, University of Stirling, UK.
pages cm
ISBN 978–0–230–28457–9
1. Transgenic organisms—Government policy. 2. Transgenic organisms—Public opinions. 3. Transgenic organisms—Environmental aspects. I. Title.
QH442.6.S74 2015
381'.41—dc23 2014029667

To Setareh

Contents

Tables

Acknowledgements

The author gratefully acknowledges the permission granted by MIT Press to re-use material that has appeared in an earlier journal article entitled 'Revisiting the Transatlantic Divergence over GMOs: Toward a Cultural-Political Analysis'. The article was published in 2012 in the journal *Global Environmental Politics*, Vol. 12, No. 4, pp. 104–124.

Abbreviations

Agbiotech	agricultural biotechnology
APHIS	Animal and Plant Health Inspection Service
ECJ	European Court of Justice
EFSA	European Food Safety Authority
EP	European Parliament
EPA	Environmental Protection Agency
FDA	Food and Drug Administration
FOET	Foundation on Economic Trends
GMO	genetically modified organism
NGO	non-governmental organisation
USDA	US Department of Agriculture
WTO	World Trade Organization

Introduction

In the autumn of 2006, a long-awaited verdict of the Dispute Panel at the World Trade Organization (WTO) was published. It concerned the divergent regulatory treatment of genetically modified organisms (GMOs) and products thereof by the US and the EU. According to some analysts, the ruling constituted a resounding victory for the US and its allies because it found against the Europeans on all major issues and condemned their failure to conform to the 1995 WTO Agreement on the Application of Sanitary and Phytosanitary Measures. The Panel criticised the EU's undue delay in approving genetically modified (GM) varieties (for import or cultivation) and the proliferation of national safeguard measures based on inadequate scientific risk assessments. Commentators also noted, however, that the Panel steered clear of the most controversial areas – such as the question of physical safety and the legality of labelling requirements – by focusing on procedural aspects and producing a relatively narrow legal opinion (Cheyne 2008). In many ways, the outcome thus mirrored a similar ruling in 1998 against the EU with regard to its ban on the use of beef hormones (rBST).

Of course, the Dispute Panel's ruling on GMOs did not end the controversy. The European Commission declared that it saw no reason to change its current regulatory framework, while simultaneously increasing the pressure on recalcitrant member states to stop using national safeguard measures. Most analysts would, in fact, trace the transatlantic controversy all the way back to the late 1980s, when a more restrictive set of European regulations was being formulated. The WTO's verdict denied the relevance of EU's 'precautionary approach' to biotechnology which the EU had managed to enshrine in the 2000 Cartagena Protocol on Biosafety (to the UN Convention on Biological Diversity) – in part because the US was not a party to the Protocol.

But the WTO's ruling has not yet led to significant regulatory convergence. As further explained in Chapter 1, there has been remarkable regulatory stability in the US since the 1980s and rapid change in the EU between 1997 and 2003, followed by a period of consolidation. In the US, authorisations of products derived from agricultural biotechnology (agbiotech) are channelled through a long-established regulatory framework. A simple 'notification procedure' is often sufficient and a lengthier 'authorisation process', which includes a full environmental assessment, can be avoided. Once approved, there are no provisions for systematic post-release oversight, as GM crops are regarded as 'substantially equivalent' to non-GM crops. By contrast, the EU's 1997 Novel Foods Regulation, which included relatively moderate provisions on GMOs, quickly became meaningless as a wave of public opposition swept through Europe and led several member state governments to resort to national bans on GM crops. A raft of new regulations emerged in the early 2000s, containing precautionary clauses on traceability and mandatory labelling, post-release monitoring, and tough thresholds for tolerable 'contamination' of non-GM products. The European Food Safety Authority (EFSA) was created in 2002 to re-establish scientific credibility and produce authoritative risk assessments of new GM products. But despite the revamped regulatory framework, only a limited number of new GM crops have been approved (for import, not cultivation), and the Commission's attempts to force member states to repeal their 'safeguard bans' have failed.

While this book does not examine the US–EU trade dispute itself, it is inspired by the persistent transatlantic divide over GMOs. It draws on a large number of existing analyses of agbiotech regulation, but in contrast to many of these writings – grounded in international law (Scherzberg 2006; Bevilacqua 2007), regulatory politics and institutionalism (Lynch and Vogel 2001; Pollack and Shaffer 2009; Sheingate 2009; Vogel 2012), political economy (Bernauer 2003; Kurzer and Cooper 2007; Falkner 2009), or critical theory (Andrée 2007) – I seek to uncover the historical and cultural origins of the transatlantic rift. While existing multi-causal accounts provide numerous crucial insights for understanding the different regulatory pathways and constellations of political and economic interest, they should be complemented by an appreciation of historically evolved structural factors, such as divergent public attitudes and cultural values/identities with relevance for GMOs. In this sense, my purpose is to revisit the existing stock of explanations and add to our understanding of regulatory divergence by developing a distinctive cultural-political analysis. Before contextualising this approach and outlining the content of the various chapters, however, a brief overview of the general debate over agbiotech is in order.

A primer on agricultural biotechnology

Since the first commercialisation of GM crops in 1996, global acreage has continuously grown at single- or double-digit rates. By the year 2013, it amounted to 175.2 million hectares in 27 different countries (around 8.16% of global cropland), although the great majority of crops (92%) were grown in only six countries (US, Brazil, Argentina, India, Canada, and China), with the US alone accounting for almost 40% of the total acreage (James 2014). The main GM crops to be planted were soya, maize, cotton, canola, and sugar beet, and most of these were engineered to withstand particular herbicides. Further efforts are underway to commercialise or expand cultivation of plants as diverse as GM trees, papayas, squash, and tomatoes as well as animals such as GM fish or pigs. The US has so far approved 196 different GM crop varieties for import and cultivation, while the EU has authorised 67 GM crops – only one of which (MON810 maize) is currently permitted for cultivation (Inghelbrecht et al. 2014). Five European countries (Spain, Portugal, Czech Republic, Slovakia, and Romania) grew GM maize in 2013, but cropland for GM maize only accounted for 1.45% (139,000 acres) of total EU maize acreage (USDA Foreign Agricultural Service 2013).

The genetic modification of plants and animals is, strictly speaking, not a new phenomenon. Ever since the beginning of agricultural civilisation, farmers have selected especially hardy or high-yielding crops. Over the past few centuries, cross-breeding of particular varieties or grafting has equally become an established method. Nonetheless, it would be easy to overstate the continuity of past activities with the modern technology of genetic engineering. Biotechnology is distinguished from other technologies dealing with inert matter by its ability to modify living organisms with the capacity for autonomous reproduction. What biologists call recombinant DNA technology, first successfully performed in 1973, allows scientists to directly manipulate the DNA of individual cells in order to change the genetic make-up of yeast, bacteria, mammalian cells, or plant cells (Manning 2000). 'Red' biotechnology focuses on the human body and medical applications, while 'green' biotechnology (agbiotech) is used to endow plants or animals with desirable traits leading to, for instance, resistance to pests or herbicides, to higher yields, or to drought resistance.

Numerous claims have been made with regard to the benefits/promises and costs/risks of GMOs. Some of the promises are as follows:

- more nutritious foods (fortified with extra minerals or vitamins);
- cheaper medicines and raw materials through 'bio-pharming' (producing vaccines or other substances in plants or animals);
- environmental benefits (reduced use of pesticides, no-till farming, and lower carbon emissions);
- major advances in crop breeding for challenging environmental conditions (salt- or drought-tolerant crops) and higher yields per acre (e.g. staple crops) or per unit of time (e.g. faster growing fish); and
- socio-economic benefits (higher farm incomes at reduced rate of labour, reduction in hunger and malnutrition).

Many of these promises have not been fully realised so far, and even modest increases in agricultural efficiency or reductions in carbon emissions have been contested. Bernauer (2003) and Black (2008) found that no unambiguous benefits were derived from GM crops. By contrast, the rate of (frequently illegal) adoption in some developing countries, such as Argentina or Pakistan, and recent studies by Barrows et al. (2014) and Brookes and Barfoot (2014) suggest modest but positive results. At the same time, a number of food contamination scandals (e.g. StarLink in 2000[1]), environmental scares (apparent toxicity to Monarch butterflies; potential contamination of native Mexican maize varieties), and occasional reports of GM crop failures have given succour to the critics of agbiotech. An international coalition of non-governmental organisations (NGOs) compiled a report to catalogue allegations of substantial risks to health, environment, and sustainable development (Navdanya International 2011).

The most prominent concerns are as follows:

- potential risks to human health (e.g. allergenicity, use of antibiotic-resistant marker genes, possibly unknown levels of toxicity, and impacts on the immune system);
- environmental impacts (creation of 'superweeds' through cross-pollination, new invasive species, non-target effects (e.g. on soil microbes or insects and farm birds), gene flow from farm crops to wild relatives and/or traditional varieties, reduction of global biodiversity as a result of intended monocultures or unintended consequences of new management practices) (Hill and Sendashonga 2006);
- socio-economic implications (domination of the global food system by a handful of multinational agribusiness firms, technological

supremacy of industrialised countries, and the threat of 'bio-piracy' to genetic resources in the developing world); and
- ethical concerns (tampering with 'nature', disregard for the intrinsic values of plant and animal species, religious food taboos, consumers' 'right to know' through labelling).

Unintended environmental consequences have indeed been identified, especially the possibility of even more efficient monocultures and incidences of cross-pollination. While there is currently not enough evidence to warrant concerns over potential impacts on human health (Barrows et al. 2014), there is still a theoretical possibility of harm (e.g. allergenicity, toxins) because of a relative dearth of long-term studies (Dona and Arvanitoyannis 2008) and because conventional risk assessment does not adequately account for scientific uncertainty (Stirling 2003; Falkner 2007).

Overall, both natural-scientific and socio-economic assessments of agbiotech tend to be interpreted according to a commentators' location on the continuum from 'promethean enthusiasm' to precautionary scepticism (Vogler and McGraw 2000). Regarding socio-economic effects, the UN's International Assessment of Agricultural Knowledge, Science and Technology for Development (IAASTD) produced a comprehensive report in 2008 that acknowledged the potential of some genetic modifications (such as drought resistance), but judged that the evidence on the benefits of GM crops had so far been patchy and occasionally contradictory. In contrast to some other environmental issue-areas – such as the threat of ozone depletion or the toxic effects of mercury – there is no genuine global consensus on the probability, scale, or acceptability of risks arising from the widespread release of GMOs. Although the science-policy nexus and the debates over potential health risks or the agricultural promise of GMOs constitute productive areas for further research, they are not the focus of this book. The aspects explored here centre on public opinion and cultural values and identities, which are areas where a significant degree of transatlantic divergence can be demonstrated and which have been closely linked to successful anti-GMO mobilisation, especially in the EU.

Understanding the transatlantic divide over agricultural biotechnology

Understanding transatlantic regulatory divergence requires a focus on the domestic politics of these two 'world regions', including longer term historical elements of socio-cultural evolution. Systematic empirical analysis of transatlantic regulatory trends over the last few decades has been performed by several scholars (e.g. Baldwin 2009; Wiener et al. 2011; Vogel 2012) and has typically built on the notion of 'risk' to explore questions of risk governance, assessment, and management. Risk has, for some time now, been of great interest to scholars of public policy (Hood et al. 2001; Smith 2004). Rather than adopting a natural-scientific and objectivist approach to risk, many analysts have highlighted the fundamentally subjective nature of risk perception by pointing to psychometric studies (Slovic 1987) and the constructivist perspective of cultural theory (Douglas and Wildavsky 1982). To analyse environmental issues, cultural theorists, such as Karl Dake (1992), concentrate on clearly delineated 'myths of nature', which are derived from broad socio-cultural dispositions (individualism, egalitarianism, hierarchism, fatalism, autonomy), and then show how these worldviews shape people's concrete risk perceptions.

Another approach, the risk society thesis pioneered by Ulrich Beck (1992), straddles subjective and objective categories of risk perception. Beck argues that the era of modernity and its technological innovations have created new and often unknown (and potentially unknowable) types of risks – such as nuclear power or GMOs – which have begun to preoccupy societies as they move from 'simple' towards 'advanced' modernity. While scholars of public policy, such as Hood et al. (2001: 5), often acknowledge the potential merit of such 'macroscopic and world-historical perspectives', they prefer to engage in detailed comparative analysis of different 'risk-regulation regimes'. Scepticism about broad historical trends or socio-cultural generalisations is also expressed by scholars examining an expansive range of transatlantic regulatory differences (Baldwin 2009; Wiener et al. 2011). By contrast, Vogel (2012) explores a limited array of public health and environmental issues and posits an underlying 'flip-flop' dynamic from precautionary to permissive regulations in the US and vice versa in the EU.

The purpose of the present book is different. It does not attempt a comprehensive transatlantic comparison across several issue-areas, although my argument about a cultural, historically evolved transatlantic divide over agbiotech regulation leans towards macro-sociological (or 'macroscopic') perspectives and lends more support to Vogel's thesis than to those questioning fundamental transatlantic differences. Furthermore, to underline the peculiar characteristics of the agbiotech controversy – as a contest over 'nature' and cultural values rather than environmental quality or public health – I do not adopt the language and categories associated with the literature on risk regulation.

First, GMOs are subject to a highly subjective understanding of risk. The importance of public opinion in shaping US and EU regulatory frameworks means that 'lay' rather than 'expert' risk assessment is often a decisive factor. And this gives rise to a distinct evaluation of agbiotech, grounded in an 'expanded vocabulary of risk that includes questions of culture, history, and ethics' (Wilkins 2001: 168). Second, the 'risk society' thesis is only partially relevant because severe hazards from agbiotech have so far not been identified. As Gaskell et al. (2004) note, both policy experts and social science researchers have often framed GMOs as a 'risk controversy', but this is not borne out by the results of EU-wide opinion surveys and focus group discussions. As long as respondents perceived no genuine benefit from GM products for consumers or society as a whole, they were reluctant to engage in any form of risk–benefit analysis.[2] Third, my own cultural, historical, and context-specific inquiry differs substantially from the formalistic quality of the categories employed by cultural theory (Sjöberg 2000). While clearly defined, ideal-type worldviews may be useful for characterising agbiotech-related attitudes of particular groups in society (e.g. farmers), cautious generalisations across many countries or regions require a more flexible language of culture, providing both for underlying commonalities and differentiation.

A cultural-political approach

Many accounts of agbiotech regulation implicitly acknowledge that culture plays some role in public opinion, political contests, and regulatory outcomes. When rendered more explicit, cultural factors manifest themselves in terms of social trust, regulatory styles or traditions, and public 'outrage'. But the long-running debate over the relationship between culture and politics is rarely addressed. When sociologist Sheila Jasanoff (2005) uses the notion of 'political culture' in her study of agbiotech regulation, she strives to 'capture the stabilities in social practices and meaning making while getting below the bland surfaces of formal politics and decision making'. And while political scientists Montpetit and Rouillard appreciate the relative neglect of culture in existing scholarship, they regard culture as a 'repertory of actions' and insist that it is 'dynamic and always contested' (2008: 927); therefore, it cannot be reduced to attitudes and values. These interventions broadly correspond to what Mishler and Pollack (2003) conceptualise as 'thin' culture. By this they mean an empirical notion of culture which is constructivist (partly chosen), endogenous (created by institutions and actions), ambivalent (a general tendency or set of attitudes with significant internal variations), dynamic (liable to rapid shifts), and individualist (best measured at the micro-level of individuals or small groups). By contrast, their notion of 'thick culture' emphasises culture's 'aggregate and holistic nature, its rootedness in history, its connectedness to society and ethnicity, its stability and resistance to change, its coherent structure as a network of meaning, its deductive character, and its exogenous nature as a determinant of both political structure and behaviour' (ibid.: 238). This is an ideal-type end point on a continuum from 'thick' to no culture – with 'thin' culture located somewhere in the middle.

The approach I adopt in this book shares some characteristics with both 'thick' and 'thin' culture. Yet, whereas most social scientists prefer the notion of 'thin' culture, and rational-choice scholars may even entirely neglect the concept, my approach is closer to the assumptions behind 'thick' culture. For example, while a rational-choice explanation may explain European resistance to agbiotech by reference to bureaucratic politics (in the European Commission), multi-level governance (with many potential veto points), and the strategies of influential interest groups, a 'thin' cultural-political account may point to the upsurge in public concerns in the late 1990s and link this to a series of high-profile food crises. By contrast, a 'thick' cultural-political account underlines the early signs of significant public concern over biotechnology (well before the mad cow crisis and heightened media interest). For the US, a 'thick' cultural-political account emphasises the difficulty of sustained public mobilisation in a context of widely shared utilitarian values and the absence of significant hazards from GMOs.

Culture, in other words, represents the context for political agency. It is 'the basis for social and political identity that affects how people line up and how they act on a wide range of matters' (Ross 1997: 42). Cultural values have to be seen as prior to the formation of interests because these are ultimately defined intersubjectively (Dyer 1996; Ross 1997). As a contextual, catalytic force, culture does not provide a direct causal explanation of political dynamics, but it offers a plausible, interpretive account of underlying structures and motivations which indirectly (and often strongly) shape political strategies and decisions. Culture should therefore be regarded as a core part of an analysis which also includes other factors derived from regulatory politics, interest group behaviour, and public mobilisation. A small number of scholars have used cultural factors to explain the transatlantic divergence over agbiotech (Krenzler and MacGregor 2000; Coburn 2005; Zurek 2007; Schurman and Munro

2010; Stephan 2012). This book draws on their insights, but none of these writers have applied a historical and macro-sociological lens to the controversy. Rather than limiting oneself to an analysis of contemporary cultural politics, it is also important to inquire into the origins of cultural values and identities which underpin the public's responses to GMOs. This may help to gauge the intensity and durability of relevant cultural elements and thus inform political strategies in support of or against GMOs.

Besides revisiting the existing scholarship on the topic, the contribution of this book thus centres on a careful consideration of the historical and cultural sources of contemporary policies. A diverse range of academic contributions – from disciplines such as history, sociology, and anthropology – provide the foundation for a more extensive cultural-political analysis. My core argument revolves around the plural meanings of 'nature' and the point that many Europeans (and far fewer Americans) perceive GMOs as a threat to one or several of these connotations, such as cultural identities or traditional livelihoods related to food and agriculture or the idea of a metaphysical 'natural order'. In Europe, agbiotech has come to represent a 'sounding board' (Torgersen et al. 2002) for contemporary anxieties about modernity, globalisation, and the decline of national identity. In various combinations, these concerns give rise to a potent moral critique of the 'unnaturalness' of GMOs, which often crowds out utilitarian risk/benefit evaluations. In the US, these dynamics are significantly weaker, even though other (relatively effective) anti-GMO narratives have recently gained prominence.

The first part of this book (Chapters 1–4) summarises the core aspects of transatlantic regulatory divergence, proposes a cultural-political approach, and performs a detailed analysis of resistance to GMOs in the US and the EU. The second part (Chapters 5–6) delves into the historical currents that inform European and American attitudes to the 'natural order' and 'unnaturalness', the ideal landscape, food cultures, and agricultural traditions.

Structure of the book

More specifically, Chapter 1 examines the regulatory frameworks for agbiotech and public opinion trends on both sides of the Atlantic and traces their evolution over the past decades.

Chapter 2 surveys the most influential explanations of the transatlantic regulatory divide. Institutionalist perspectives deliver comparative analyses of political and administrative systems, regulatory styles, and historical, evolutionary logics that lie behind regulatory stability and change. Political-economic perspectives offer both actor- and society-centred explanations of regulatory dynamics, giving due consideration to various interest groups as well as public opinion. In combination, these approaches provide highly persuasive, multi-causal accounts of transatlantic regulatory divergence. However, if 'public outrage' in the EU and relative tolerance in the US constitute crucial explanatory factors, then more deserves to be said about the roots, durability, and catalytic quality of public opinion. The second part of the chapter thus reviews a range of aspects associated with public attitudes (scientific literacy, media influence, social trust) and concludes that a blend of ethics and morality provides key cultural elements that can be integrated into a broader, historically informed, political analysis.

Chapter 3 explores the concepts of culture and civilisation and suggests that viewing culture as context allows for a partial reconciliation with the study of politics. One way of operationalising the notion of cultural context is to gauge the cultural resonance of particular discourses. This may be done to verify the centrality of associated cultural values and/or identities, but political mobilisation (e.g. by NGOs) is needed to turn them into salient factors that outweigh more familiar utilitarian considerations and help to shape the political agenda. The most relevant cultural attitudes underlying public anxiety about agbiotech are associated with ideas of 'naturalness' and the 'natural order'. These are perceived to be under threat from a technological quantum leap with unpredictable moral and material implications. GMOs are regarded as a prime instance of 'capitalist modernity' with its seemingly unstoppable drive towards rationalisation, efficiency, and commodification of culturally significant goods or practices rooted in the lifeworld. This broad category of reactions can be observed across many different European countries.

The cultural politics of agbiotech is analysed in more depth in Chapter 4. While NGO mobilisation in Europe has been vital in building up popular pressure for regulatory stringency, the success of these campaigns has been related not only to political opportunity structures but also to the ability of anti-GMO discourses to 'activate' widespread and pre-existing, culture-based anxieties. American NGOs have tried to replicate European success, but US activists have faced less favourable political opportunity structures and a powerful pro-agbiotech coalition. Predominantly utilitarian discourses have not resonated or cut across political boundaries to the same degree. This has only recently begun to change with a renewed emphasis on 'consumer sovereignty' as the overarching moral and political mission. The differences in activists' narratives and the broader cultural context are mirrored in the responses of farmers' organisations and the strategies of corporations: optimism in the US (with exceptions) contrasts with cautious incrementalism or even apathy in Europe. However, besides drawing attention to the importance of cultural context, the relative effectiveness of activists' discourses remains a limited indicator of cultural-political dynamics. Only historical analysis can show how cultural attitudes and identity politics have followed a distinctive trajectory in Europe, while creating a more hospitable context for GMOs in the US.

Chapter 5 draws on the work of environmental and cultural historians and summarises the role of 'nature' and the 'un/natural' in America (since the first English settlements) and Europe (since the Middle Ages). A pattern begins to emerge early on. Whereas European relations to the natural environment were marked by material interdependence and spiritual precepts, American settlers shook off their inherited attitudes relatively swiftly and took up the challenge of mastering a 'wild' continent. These dynamics gave rise to divergent 'civilisational dispositions' – the tendency of perceiving relations between humanity and nature as either 'interactive' or 'bifurcated'. The second part of the chapter tracks the aestheticisation and nationalisation of nature. The former is closely associated with the Romantic movement and the latter with the formation of national cultural identities. The core transatlantic difference relates to 'bifurcationist' America shifting from a European tradition of pastoral beauty towards the sublimity of wilderness, albeit alongside continuing exploitation of nature. In the US, nature has taken on a meaning of magnificent grandeur and pure truth, while in 'interactive' Europe it has remained mired in compromises. The pastoral ideal of landscape and nature, which includes socio-economic and aesthetic elements, has remained strong in Europe. Throughout history, it has fuelled important movements of resistance against modern industrial or technological change. Nationalism and cultural identities have frequently drawn on images of the 'middle landscape' – on artisanal and agricultural livelihoods and their associated humanised environments. The arrival of new technologies is not a problem per se, but much depends on the cultural meanings attached to them.

Building on the historical comparison, Chapter 6 further develops the linkage between cultural attitudes and European public resistance to agbiotech. Drawing on sociological perspectives on modernity, I highlight the similarity of contemporary (late modern) European opposition to GMOs with pre-modern critiques of modernisation. Both types of movements expressed their resistance through steadfast moral defiance, boycotts, and appeals to tradition, identity, and nature. European societies are part of an 'interactive', 'agri-cultural' civilisation which recognises associations with the 'natural' in its food products and mixed cultural landscapes. By contrast, American society is portrayed as a 'bifurcationist', modernist cultural context in which the values of efficiency, simplicity, and abundance tend to crowd out neo-traditionalist concerns about stable cultural identities, the middle landscape, and the culinary heritage. Notwithstanding internal diversity in both world regions, American and European societies have historically moved along distinct cultural trajectories. These are not diametrically opposed and do not directly cause regulatory outcomes. But they are marked by significant differences and have a powerful influence on the politics of agbiotech in the US and the EU.

1
Overview of Regulatory Frameworks and Public Opinion

Introduction

This chapter lays the groundwork for a cultural-political analysis. The regulatory pathways described here seem to confirm the assumptions of historical path dependency in which initial political decisions decisively shape the interests of rational economic actors and structure the field of political possibilities (Pollack and Shaffer 2009). Such accounts are plausible, but they tend to underestimate other factors influencing the regulatory trajectory, especially the role of public opinion and of bureaucratic politics. In Europe, the latter shaped the early framing of biotechnology as an environmental question – to be supervised by the EU's Environment Directorate-General (Patterson 2000). My main focus, however, is squarely on the public mood and the amount of political leeway it offers, particularly once public opinion became subject to regular surveys in the 1990s. While, in the EU, bureaucratic politics, industrial policy priorities, and major economic interests were drifting towards a US-style regulatory framework by the mid-1990s, this developmental path was thwarted by the anti-GMO mobilisation of citizens and consumers. The precautionary logic of the initial framework from 1990 has been preserved, even if greater centralisation at the European level implied a possible mechanism for modest regulatory softening and more technocratic policy-making.

The European story thus contradicts simple models of unproblematic technological diffusion. In the US, on the other hand, agbiotech seemed to follow the expected pathway of technological innovation and modern socio-economic progress. Here, both scientific organisations and industrial policy-makers have long supported scientific research and commercialisation through permissive regulations, while public opinion has remained accommodating (Gaskell et al. 2002: 351). This settled situation has only begun to change over the last few years. The European pro-GMO camp regularly points across the Atlantic and urges policy-makers to follow the exemplary US path of innovation to bolster Europe's economies against competition in a rapidly globalising world. In this chapter, I largely confirm this fundamental transatlantic divergence by summarising the regulatory history of the two regions, highlighting core regulatory principles, and gauging prevalent trends in public opinion.

A short history of agbiotech regulation in the US

American scientists can be regarded as 'first-movers' in both scientific and regulatory aspects of biotechnology. After the discovery of recombinant DNA in 1973, intense debates began within the wider scientific community. In February 1975, a historic meeting was convened in Asilomar to discuss the ramifications of the scientific breakthrough – a cornucopia (or, alternatively, Pandora's Box) in the eyes of many of the 140 participating scientists and lawyers. Over 30 years later, many still hailed this first attempt at responsible scientific self-regulation as a milestone, a genuine 'Woodstock of molecular biology' (Barinaga 2000). But Asilomar's legacy is contentious and the dynamic of self-regulation it helped trigger also dampened regulatory activity in subsequent decades. Jeremy Rifkin of the Foundation on Economic Trends (FOET) claimed that the short-lived moratorium on genetic research – intended to provide the space for a consensual regulatory solution – was ultimately motivated by 'issues of personal and institutional liability' for potential 'bio-hazards' and not by caution over health or environmental consequences (Rifkin 1999: xi). The technical measures of biological containment proved to be a powerful, soothing gesture towards American law-makers and the public at large, signalling that scientists were able to 'manage their own business'. Crucially, the long-term effect was that 'the economic, social, political, military, ethical, and future ecological issues largely dropped from public view' (Regal 1999).

This impact has been visible in a series of institutional reforms over the years. With Asilomar's effect of public reassurance, genetic technology was to be monitored by a Federal Agency, the DNA Advisory Committee of the National Institutes of Health. The arrangement of arms-length supervision was maintained for approximately nine years. It did not end due to public debate or intra-scientific dispute but through a successful lawsuit brought by Rifkin's FOET. Without this legal action, it is doubtful whether there would have been any dedicated regulation for GMO releases into the environment (Toke 2004: 110). In response, the White House Office of Science and Technology Policy (OSTP) finalised the 'Coordinated Framework for Regulation of Biotechnology' in 1986 whose central tenet was that GM products would continue to be regulated 'according to their characteristics and unique features rather than their production methods' (Becker and Cowan 2006: 6). This framework continues to provide the basis of the regulatory system. Its three lead agencies are the Animal and Plant Health Inspection Service (APHIS) of the US Department of Agriculture (USDA), the Food and Drug Administration (FDA), and the Environmental Protection Agency (EPA) (Table 1.1).

APHIS regulates field tests and inter-state shipments of GM plants that could become plant pests under the Federal Plant Protection Act. Since 1993, the overwhelming majority of authorisations for such 'regulated articles' can be obtained through a relatively simple 'notification process'. In this procedure, which serves to expedite authorisations, a notification letter with a brief description of the genetic modification is usually sufficient. The overwhelming majority of applications fall into this category, but some – including plants producing pharmaceuticals – require a special permit. Under the 'permit process' companies have to follow relatively strict guidance from APHIS to commence field tests or to import and transport GM plants. Regardless of which procedure is followed, the next step is to apply for 'non-regulated' status in order to avoid any further formal oversight. APHIS performs a formal environmental assessment and has to allow for a period of public comment before making its final decision. In October 2008,

Table 1.1 Overview of agency responsibilities

Agency	Products regulated	Reviews for safety
USDA (APHIS)	Plant pests, plants, veterinary biologics	Safe to grow
FDA	Food, feed, food additives, veterinary drugs	Safe to eat
EPA	Microbial/plant pesticides, new uses of existing pesticides, novel microorganisms	Safe for the environment Safety for a new use of a companion herbicide

Source: Adapted from Vogt (2001).

the agency proposed a number of regulatory changes, including the discontinuation of the notification procedures and the introduction of a new petition procedure which would offer conditional exemption from permit requirements (Cowan 2010). But no final decision has been made so far.

Operating under the principle of 'substantial equivalence' between GM and non-GM foods, the FDA draws on its powers under the Federal Food, Drug and Cosmetic Act and the Public Health Service Act to regulate food additives. Only additives that are 'generally recognised as safe' (GRAS) do not need pre-market approval. Following an unsuccessful lawsuit by a coalition of NGOs in 2000, the FDA began to encourage developers of GMOs to engage in voluntary pre-market consultation. In June 2006, the FDA went further by promoting consultation in the pre-development stage of GMO research in order to better identify new proteins or other additives that might later pose safety risks such as allergenicity (Becker and Cowan 2006). In January 2009, the FDA announced that, unlike plant-based GM foods, GM animals and derived products would require FDA pre-market approval.

EPA regulation relies on the Federal Insecticide, Fungicide, and Rodenticide Act (FIFRA) and its provisions for the registration of new pesticides. This includes GM plants that produce their own pesticides. Companies wishing to use a new pesticide in field trials must provide a substantial amount of information about the product's effectiveness, as well as a broad range of data about product chemistry, toxicology, and its effect on non-target species (Belson 2000). Following the StarLink scandal in 2000, in which a GM corn approved only for animal feed found its way into taco shells, the EPA stated that it is highly unlikely to authorise a GM plant in the future that has not been cleared for human food (Vogt 2001: 17). Implicit in this announcement is the recognition that coexistence of GM and non-GM plants without routine cross-contamination would be extremely difficult. In practice, however, the EPA's authority to mandate restrictions and require post-approval monitoring and reporting is hampered by resource constraints, making it dependent on the help of state-level environment agencies and industry self-reporting.

The US regulatory system is also distinctive due to its underlying policy assumptions. First, the notion of 'substantial equivalence' implies that GM foods and crops are not properly 'novel' or pose special risks but can be dealt with by using existing regulatory routines while allowing for some limited adaptations. Second, this confidence in the 'neutrality' of the new technology reinforces the commitment to product- rather than process-based regulations, in which not the method of production but the properties of the final product are scrutinised. Third, such underlying beliefs exist in a recursive relationship with a more general 'free market approach' of US regulatory policy-making, whereby it is the government's duty to stimulate societal progress by 'avoiding undue burdens for technological advance' while ensuring the safety of new products (Toke 2004: 111).

Largely fashioned in the 1980s, the US regulatory framework has a facilitating quality and makes use of many voluntary elements of consultation. This more lenient approach is complemented by the threat of litigation – both against individual companies and federal regulators themselves. As mentioned above, legal action rather than broad-based popular pressure gave rise to the formulation of a systematic framework in the first place (Hallman 2000). On the whole, however, both the insistence on 'sound science' and the professed even-handedness of regulatory measures are misleading. Core scientific principles of US regulations, such as 'substantial equivalence', cannot be understood outside their political, legal, and cultural context (Pelletier 2006), and the divergence between proposed regulations and actual decisions shows that regulatory outcomes are biased in favour of food producers and retailers (Haniotis 2001). This, in turn, is not simply a result of lobbying power. The oft-criticised 'revolving-door' migration among regulators and industry lobbyists is no mere political flaw but also the expression of a wider belief system that produces a far-reaching consensus between the two groups (Toke 2004: 113): to them, agbiotech products are broadly benign, progressive, and offer impressive economic opportunities. Available data on US public opinion suggest that a largely positive reading of technological progress is mirrored in the general population, albeit with weaker confidence in its safety and latent concerns about both physical and ethical ramifications.

US public opinion since 1987

Surveys have been widely used in the US and elsewhere because they offer a glimpse of public opinion in a quantified format, providing easily understandable material for policy-makers and the media. Increasingly, however, commentators are treating attitudinal surveys with caution. Different surveys often come to divergent conclusions despite querying the same public; too much evidently depends on how questions are framed, on what terminology is used or on what information is supplied. Too often, therefore, policy-makers base their decisions on researchers' 'perceptions of public perceptions' (Cormick 2005: 227) rather than on an assessment of a broader range of available data. Davison et al. (1997: 318) go as far as charging opinion polls with actively constructing and rationalising public debate into quantified, simplistic distortions.

According to them, broader concerns about the underlying political economy and cultural significance of biotechnology are concealed by a narrow focus on professional-utilitarian ethics and an emphasis on consumers rather than citizens.

Notwithstanding this critique, a preliminary summary of US and EU public opinion necessarily has to rely on the available data. The main picture emerging from US surveys is that the American public appears to be in a 'state of schizophrenic tension, with the majority of people simultaneously expressing optimism about the potential benefits of GM technology and concern about the unforeseen consequences of its use' (Schilling et al. 2002: 8). Overall, public acceptance has followed a long, gradual sinusoidal curve over the past two decades, displaying a peak of concern around 2001. Until 1997, there was little consumer concern in the US and between two-thirds and three-quarters of consumers seemed willing to accept GM foods (Hoban 1997: 232). This rosy picture changed when agbiotech first emerged as a genuine item on the public's agenda – partly driven by persistent controversy in Europe. US media coverage increased and took on a more critical orientation (Shanahan et al. 2001). There appears to have been a peak in both concern and awareness around 2000–2001 when only half the US population still believed in substantial benefits of agbiotech over the next 20 years (Priest 2000: 939), and general public awareness reached 53% (Hoban 2004). This phenomenon correlates with the activities of an NGO coalition between 1999 and 2001, which is discussed in Chapter 4. By 2006, opposition to GM foods had declined again somewhat and was estimated at 34% by one group of scholars (Ganière et al. 2006: 146). By 2012, in a survey devised by the International Food Information Council, 38% of respondents were somewhat or very favourable towards plant biotechnology, while 20% were somewhat or very unfavourable (IFIC 2012).

Among researchers on US public opinion, there is general agreement on a number of elements: low awareness and knowledge of food biotechnology as well as a clear distinction between plant and animal biotechnology. Hallman et al. (2004) write that – despite considerable media coverage of agbiotech issues in previous years – '[m]ost Americans have heard or read little about it, are not aware of its prevalence in their lives, and are confused as to which type of GM products are available'. Virtual consensus exists on the fact that Americans much more readily approve of plant-based GMOs than of animal-based GM technology. But scholars diverge on whether survey data is ultimately sufficient to reach adequate conclusions about public opinion. Schilling et al.'s (2002) notion of 'schizophrenia', which can be paraphrased as inconsistency, is backed by empirical data. The first ever poll on public sentiment, commissioned by the US Office of Technology Assessment in 1987, shows the kind of ambiguity that can still be observed today. While 66% of respondents believed that genetic engineering would improve the lives of Americans, 52% were anxious about the dangers to people or the environment (Ezzell 1987). Furthermore, respondents were much more tolerant of technological risks if these were presented in quantified rather than abstract terms, and ethical concerns were often overruled when faced with real-life medical or environmental benefits (Ezzell 1987; Hallman et al. 2003). The unstable nature of survey responses led some to describe them as 'often equivocal and highly malleable' or 'uncrystallised' (Hallman et al. 2004). Hallman and Hebden (2005: 241) concluded that, on the whole, opinions are 'weakly held' and 'poorly formed'. Opinions are malleable and will often change when presented with additional information about GM foods such as benefits and risks regarding public health or the environment (Fink and Rodemeyer 2007). In essence, therefore, public opinion on agbiotech remains somewhat 'up for grabs' (Pollack and Shaffer 2009: 267).

Another important question concerns the near-universal demand for mandatory labelling of GM products. Around 80% of consumers – rising to 93% by 2013 (Kopicki 2013) – want special labels on food products, but only 53% would consequently buy less GM produce (Fink and Rodemeyer 2007). Hallman et al. (2004: 11) assert that this demand is only part of a whole raft of calls for more labelling –the use of pesticides claiming the top of the hierarchy and GMOs coming second. In 'open' questions about food labelling (which do not give respondents pre-selected options), a mere 1–3% of respondents mentioned GMOs as an important issue (Hoban 2004; IFIC 2010; 2012), rising to 7% amid the latest wave of activist campaigning (Hallman et al. 2013). This disjuncture may be best explained by the pervasive leitmotif of 'consumer sovereignty'. More information is always welcomed by consumers because it allows them to enhance their control over buying decisions.

Given a degree of scepticism about the adequacy of surveys for measuring public opinion, studies conducted by the non-partisan Pew Initiative on Biotechnology in the mid-2000s also relied on focus groups to corroborate polling data (Fink and Rodemeyer 2007). Taking their findings into account, it can be concluded that – amid a generally favourable climate for agbiotech and technological innovation – opposition reaches beyond a narrow 'fringe' of the US population, even if many of the concerns voiced are rather dynamic and malleable in nature (see also Priest 2000: 940).

Pressures for US regulatory reform

Public opinion, associated campaigning, and legal challenges by civil society organisations constitute only one major driver of regulatory reform in the US. Much of the discussion in Chapters 2 and 4 will be devoted to the above factors, but other aspects also deserve some attention. Although the regulatory landscape has been stable for many years, more recently economic and scientific-regulatory pressures have begun to mount. Political elites and regulatory agencies are now hesitantly responding to ever louder, if conflicting, calls for regulatory adaptation, triggering a process that some have dubbed 'change without reform' (Pollack and Shaffer 2009: 273).

Economic challenges to the development and widespread planting of new GM varieties have come through legal activism in the US and the desire to preserve agricultural exports to major foreign markets. First, domestic opponents of GMOs (NGOs, organic and small-scale farmers), who fear the extensive contamination of their produce and the monopolisation of the seed market, have launched a series of lawsuits. Coalitions of activist organisations, often spearheaded by the Center for Food Safety, have sought bans on the planting of certain GM crops, such as alfalfa and sugar beets, as well as on field trials of GM eucalyptus trees. In the case of GM alfalfa, opponents claimed that conventional and organic crops could easily be contaminated and that this would have a negative economic impact. They also raised the possibility that increased herbicide use on GM alfalfa could lead to the emergence of resistant 'superweeds'. The plaintiffs won their court case in 2007. The ruling effectively imposed a temporary ban and required the USDA to conduct a full review. USDA's detailed report was published in December 2010, outlining the options which included a laissez-faire approach and another model based on minimum separation distances (5 miles) between GM and non-GM crops. After drawn-out deliberations, in early 2011 Agriculture Secretary Vilsack sided with the biotech companies and most major farming organisations. Declaring that farmers must have the choice of planting GM alfalfa, he endorsed a laissez-faire approach, and USDA issued a second approval of GM alfalfa without restrictions (Gillam and Doering 2011).

The case of GM sugar beets has certain parallels with the controversy over GM alfalfa. But the case acquired serious economic significance for many farmers due to the fact that GM sugar beets accounted for about 95% of all varieties sold and represented roughly half of the nation's sugar supply. In August 2010, judges ordered a ban on the planting of GM beets. Yet, by November 2010, USDA had sidestepped the ruling by re-approving the plants with strict guidelines until the completion of its environmental impact assessment. In both the alfalfa and the sugar beet cases, as well as in the eucalyptus case (which was unsuccessful in the courts), activists attempted to destabilise the domestic seed market for particular GM crops by prompting temporary bans and detailed environmental reviews. Regarding alfalfa, they also sought to protect the genetic composition and potential marketing advantage of conventional and organic varieties. Moreover, for other GM crops in the pipeline – such as GM corn, potatoes, wheat – and for the FDA's pending decision on GM salmon, such legal uncertainties are likely to translate into more testing and further delays in development and approval processes, which ultimately impacts negatively on overall profitability.

This trend of increasing economic risks comes on top of existing challenges with regard to foreign markets for US agricultural produce. The process of 'commercial adaptation' (Pollack and Shaffer 2009: 263ff.) that US activists are hoping to accelerate has long been a reality for some commodities. Not only would key markets, such as the EU and Japan, potentially reject imports of novel GM crops but also the inability of the US supply chain to reliably separate GM and non-GM produce has already created several incidents with major financial implications. For instance, in 2001 Japan recalled all products made with GM potato starch. And in the most prominent case so far – concerning an experimental rice (LibertyLink 601) developed by Bayer CropScience which in 2006 found its way into conventional rice exports – Japan, the EU, and Mexico temporarily blocked all rice imports from the US. In July 2011, Bayer agreed to pay affected US rice farmers US$750 million in compensation. In response, in 2007 the Biotechnology Industry Organization (BIO) established the Product Launch Stewardship Policy, a form of voluntary self-regulation designed to coordinate the commercialisation of GM crops – with particular attention to foreign regulatory conditions. However, in May 2013 a similar dynamic once again unfolded when a GM wheat variety (which was not approved for human consumption) was detected in a single field in Oregon, raising concerns over mismanagement by the developers (Monsanto) or even deliberate 'sabotage' by anti-GMO campaigners. In sum, economic pressures have strongly shaped the activities of biotech companies and led to a focus on GM crops with widespread commercial acceptance, especially those used as animal feed and for industrial purposes, such as maize, cotton, canola, and soybeans (ibid.: 266).

The other major pressure for regulatory reform arises from the recognition that the US regulatory system is outdated and not sufficiently comprehensive to keep up with the latest developments in biotechnology. The US had long been the only significant exporter of GM produce and, thus, USDA never developed a dedicated import control policy that could serve to identify shipments of unapproved GMOs, such as GM rice from China (Melvin 2009). This perception of regulatory inadequacy is compounded by increasing complaints from agricultural scientists about independent testing of GMOs being made impossible by corporate concerns over patents and confidential information. Even some of the champions of agbiotech, such as former special government advisor on science and technology, Nina Fedoroff, acknowledge that these gaps undermine the government's claim to sound, science-based regulation and do not help to enhance public confidence in agbiotech products (Gillam 2010). Furthermore, some GM inventions are likely to escape regulatory scrutiny altogether, unless the remit of the relevant agencies is broadened (Nature News 2013). For example, in 2011 the company Scotts Miracle-Gro developed a GM version of Kentucky bluegrass, commonly used for lawns, which promises reduced maintenance efforts. The novel genetic technique that was employed did not require the use of viruses or bacteria and, consequently, USDA declared that the plant did not fall within its remit. The GM bluegrass, as well as a string of GM innovations in the pipeline, would therefore remain unregulated and could freely enter the market (Nature 2011).

The combination of scientific uncertainty and inadequate regulatory coverage has led to increasingly loud calls for regulatory reform. According to Brown (2011), a framework for a new, more comprehensive biotechnology law 'has been sitting quietly on the shelf' for over a decade. This framework would exempt no GM organism from scientific scrutiny, it would increase transparency, and it would give much greater weight than previously to potential environmental impacts of new agbiotech releases. But in what represents a classic illustration of inter-agency struggles over regulatory 'turf', the new legislation, which was destined to endow the EPA with lead responsibility, was opposed by both USDA and FDA. While the general debate may have been reignited, no resolution is currently in sight.

In conclusion, while the US regulatory landscape is no longer as settled as it was throughout much of the 1990s and early 2000s, its limited scope and stringency provide a marked contrast to EU agbiotech regulation. Europe's regulatory framework reflects the heightened saliency of the topic in the European public sphere and provides a first indication of a profound transatlantic divide.

A short history of agbiotech regulation in the EU

The early period of regulation: From the 1970s to EU directive 90/220

The early phase of European biotechnology research was not altogether different from the American experience (Table 1.2). European policy-makers were inspired by the example of American self-regulation and sought to protect the scientific community from external regulatory interference. The first 17 years after the discovery of recombinant DNA were governed by a desire to allow biotechnology research to flourish (Grabner et al. 2001: 16). As the 1980s witnessed a resurgence of economic priorities and supply-side policy-making, the field of biotechnology assumed the status of a promising technological revolution that could boost European competitiveness and yield significant economic benefits (Gottweis 1998). Considerable amounts of capital, both public and private, were invested in the burgeoning sector. The detractors of biotechnology still relied on an ethical discourse that attacked the industrialist commodification of life for profit and 'red' (human, medical) biotechnology came in for the strongest condemnation (Torgersen et al. 2002). The 1970s and 1980s were nevertheless a time when biotechnology regulation exhibited a thoroughly traditional and technocratic quality. European countries dealt in different ways with internal opposition – responses ranging from regulatory elitism to systematic public involvement – and, by and large, they succeeded in appeasing societal concerns. But renewed pressure for regulation was coming from other quarters: the European Single Market Programme and its philosophy of harmonisation were contrary to the increasing heterogeneity of national regulatory frameworks. In combination with the rapid development of genetic technologies, the general trend towards harmonisation created strong incentives for greater European coordination.

The early battles over EU-level legislation were largely waged in the corridors of the Brussels bureaucracy. The final product of the political wrangling among the Commission's Directorates-General (DGs) was EU Directive 90/220. Its primary purpose was to address the broader environmental aspects of agbiotech. This focus was, to a large extent, both the result of and the reason for DG Environment taking the

Table 1.2 Core aspects of EU biotechnology regulation

Year	Event	Main content
1984	Commission forms biotech steering committee	
1986 (November)	Commission Report 'A Community Framework for the Regulation of Biotechnology'	– First defines the rationale for a European regulatory regime
1990	Directives 90/219 and 90/220	– Environmental precaution and process-based assessment – 'Safety clause' allows national governments to enact temporary bans on GMO releases
1996 (March)	Beginning of crisis over mad cow disease (BSE)	
1997	Novel Foods Regulation (258/97)	– Replaces centuries-old food regulation based on tradition and experience – Specific risk assessments based on principle of substantial equivalence – Simple notification procedure for GM foodstuffs not containing GMOs in finished product
1997 (March)	Commission approves sale of GM maize; three member states invoke safeguard clause	
1998 (October)	Start of de facto moratorium on approval of new GM varieties	
2000 (January)	White Paper on the Precautionary Principle	
	Cartagena Protocol on Biosafety Adopted	
2001 (March)	Directive 2001/18 replaces 90/220 on deliberate release	– Enshrines Precautionary Principle and detailed environmental risk assessment – Stipulates respect for ethical principles and mandatory public consultation – Limits all approval to an initial ten-year period

2002 (January)	Establishment of European Food Safety Authority (EFSA)	– Privileged scientific arbiter and adviser to the Commission, but its role is contested by member states' own scientific institutions
2003 (May)	US launches WTO complaint over EU regulations	
2003 (July)	Commission issues 'Coexistence Recommendation' 2003/556/EC	
2004 (April)	Regulations 1829/2003 (GM food and feed) and 1830/2003 (labelling and traceability)	– Breaks with assumption of 'substantial equivalence' – More detailed testing procedures to establish food safety of GM products – 'Contamination clause' foresees national legislation for 'coexistence' of agricultural systems
2004 (May)	Commission ends moratorium with approval of Bt-11 maize	
2006 (September)	Final report of WTO Dispute Panel	– Upholds most procedural complaints, but steers clear of substantive issues such as scientific uncertainty
2008 (February)	France proposes a review of the EU's GMO approval process	– France and other member states make the case for taking more account of socio-economic factors and giving countries more regulatory flexibility
2009 (April)	Second Coexistence Report is published	– 15 member states have adopted relevant legislation
2010 (March)	Commission authorises BASF's GM *Amflora* potato (destined for industrial starch) for cultivation	– Destined for use as industrial starch – Only the second GM crop to be cleared for cultivation in the EU

Table 1.2 (Continued)

Year	Event	Main content
2010 (July)	Commission puts forward a new 'Cultivation Package', including a communication (COM 380 final), a proposal (COM 375 final) and a recommendation (OJ C200/1)	– Proposes partial renationalisation of cultivation authorisations in return for stronger commitment to EU framework for risk assessment – Provides a more flexible approach to the management of coexistence
2011 (June)	Commission adopts Regulation 619/2011 harmonising controls for non-authorised GM material in feed	– Addresses concerns of livestock farmers and shipping companies over 'contaminated' feedstock – Feed imports may now contain 0.1% of unauthorised GM crops, but only if approved elsewhere and if an authorisation procedure is pending with EFSA
2012 (June)	Press reports indicate that Commission plans to end its zero tolerance stance on unauthorised traces of GMOs in food shipments	
2013 (September)	European Court of Justice (ECJ) rules in favour of Pioneer Hi-Bred International's request to lift the suspension on the approval of Bt-1507 maize	– EFSA had five times declared the GM crop to be safe – ECJ decided that the Commission had no reason to delay the authorisation process (Reg. 1829/2003 states that max. period after risk assessment is three months) – Forces the Commission to table a request for authorisation in the next Council meeting
2013 (December)	ECJ revokes authorisation of GM *Amflora* potato due to procedural failings	
2014 (January)	Commission indicates that it will not seek approval of further GM crops before reaching an agreement on regulatory reform	

2014 (March)	Greek presidency puts forward a compromise proposal	
2014 (June)	European Council reaches political agreement on new regulations for GMO cultivation	– Compromise agreement offers more flexibility for national cultivation bans, but maintains centralised EU system for approval of new varieties – Agreement by European Parliament is also required; adoption not expected before 2015

Source: Adapted from Pollack and Shaffer (2010).

lead role. Having emerged victorious from the power struggle with a more laissez-faire DG Science, DG Environment enshrined the logic of a process-based regulatory framework. The combined message of Directives 90/220 (on deliberate release of GMOs) and 90/219 (on contained use) was one of environmental precaution and its underlying philosophy was squarely opposed to US product-based regulations. The fundamental difference, comments Toke (2004: 158), was rooted in a negative answer to the question: 'Are GM plants "natural?"' In Europe, both field trials and commercialisation of GM plants now required explicit regulatory consent. Two further important features of Directive 90/220, however, indicated the difficulty of this first attempt at harmonisation. Opposition stemmed not so much from non-state actors as from national governments. Public controversy continued to take place mostly at national rather than European levels, although its intensity was in gradual decline in the early 1990s. The biotechnology industry accepted the need for a regulatory framework for both economic reasons and the soothing effect on public opinion.

The economic logic of the Single Market was readily acknowledged by national negotiators, but at the same time 'regulatory styles and the history of the debate across Europe varied profoundly' (Grabner et al. 2001: 17). Commandeur et al. (1996) identified the root causes of this challenge as (1) different understandings of risk and (2) different interpretations of environmental impacts. Many (especially southern) member states had no dedicated legislation for biotechnology and the various scientific arguments put forward also displayed a conflict between narrow, 'science-based', and broader 'social' approaches to the regulation of new technologies. Denmark, Sweden, Finland, and Austria publicly advocated the inclusion of indirect environmental impacts and recognition of the wider consequences of agbiotech, such as socio-economic questions and the goal of sustainable agriculture. The UK and the Netherlands were most outspoken in the camp of those wanting only direct ecological effects to be included.

The conflict over the content of the harmonised regulatory framework led to its supranational character being watered down and complemented by intergovernmental ingredients. Article 16 of Directive 90/220, the so-called safety clause, accords considerable role to member states by allowing them to restrict or prohibit a GM plant or product if they have 'justifiable reasons', such as additional scientific evidence about its potential impact on public health or the environment. The intuition behind this was that solid scientific assessments and emerging regulatory routines would, over time, make recourse to this tool unnecessary. However, this turned out to be a major misapprehension. Directive 90/220 furthermore allowed for a Regulatory Committee of national experts (Article 21) in case of disagreement over a proposed authorisation. Due to the need for unanimity, this committee never amounted to more than a discussion group (Toke 2004: 158). In summary, perhaps the most decisive result of this particular EU directive was not the modest 'disciplining' of national-level authorities but the foundation of an institutionalised regulatory pathway – displaying aspects of path dependency as discussed in Chapter 2. The new trajectory of political oversight and risk assessment reduced the likelihood of ever returning to a US-style, 'science-based' system of monitored self-regulation (Bernauer 2003: 81).

The 1990s: Underlying trends and the 'Years of Controversy'

Much of the debate over agbiotech ebbed away in subsequent years, retreating to the realm of institutional and bureaucratic politics or returning to the national sphere. As social scientists began to regularly use opinion polls to measure the public mood, it transpired that a significant level of concern existed and was actually growing throughout the early 1990s. Though public awareness stood only at around 50% in 1991, 58% of European respondents perceived some risk in genetic engineering and considered its application harmful, predominantly to farm animals (67%), food (62%), and humans (61%) (Zechendorf 1994). By 1993, about 85% of Europeans were calling for stricter regulations on genetic engineering and the overall endorsement of the technology fell from 54% in 1991 to 48% in 1993 (ibid.: 874). Looking at these early indicators of public sentiment, it can indeed be assumed that 'the seeds of rejection had already been sown in the early 1990s' (Grabner et al. 2001: 23). The 'bifurcation' of attitudes vis-à-vis 'red' biotechnology and agricultural ('green') biotechnology was also taking shape in those years. While new medical treatments and drugs were largely welcomed, 'green' agbiotech applications met with an ambivalent and frequently apprehensive response (Torgersen et al. 2002: 59).

Scientific progress in biotechnology proceeded apace during the 1990s, but commercialisation was never going to be a straightforward task. The quest for patents and marketing expertise encouraged intra-industry cooperation and produced highly contingent trajectories of product development. In the US, Calgene's *Flavr Savr* GM tomatoes, engineered for slower ripening and flavour retention, foundered mainly due to strategic economic inadequacies, while Zeneca's GM tomato puree was successfully marketed in the UK – only to fall victim to a wave of public anti-GMO mobilisation in the late 1990s (Harvey et al. 2002). The 'watershed years' of 1996–1997 (Gaskell and Bauer 2001) shattered the illusion of permissive tranquillity. Retrospectively, however, the events of the late 1990s are more readily comprehensible. The Eurobarometer survey in 1996 conjures up the image of a 'groundswell' of popular opposition, of 'people [...] for whom assurances about the absence of scientific risks would have been unlikely to alleviate their concerns' because they had become increasingly worried about the 'usefulness, risks and moral acceptability of GM foods' (Gaskell et al. 2002: 356).

Public outrage, consumer boycotts, and NGO-led mobilisation became the defining features of the European politics of biotechnology in the 'post-watershed' years. However, before public opposition had reached the echelon of top-level decision-makers, another two eventful years had to elapse. The 'trigger' episodes of the anti-GMO backlash can only be adequately understood within their own particular historical context and by drawing on the political constructions and explanations offered by the media and countless other actors. Two key events in 1996–1997 were no incontrovertible signals per se (Toke 2004: 144) until specific interpretations and narratives captured the public's imagination and ushered in a new awareness about the ongoing technological transformation of agriculture and medicine. Many parts of Europe were most affected by the arrival (in the autumn of 1996) of the first US shipments of GM soya to French ports, accompanied by Greenpeace's media-savvy protests. For much of southern Europe, on the other hand, a remarkable turnaround in public mood came only after the birth of the first cloned mammal, Dolly the sheep, was announced in February 1997.

Italy and France, for instance, only began to actively scrutinise GM food and crops after this 'wake-up call', while established biotech critics in countries such as Austria and Denmark used the GMO shipment event to raise fundamental questions about existing agricultural policies and encourage a wider debate on genetic technologies. It is worth clarifying this dynamic by briefly examining both a 'northern' and a 'southern' country during the 'watershed years'.

A 'late-comer' with no significant economic, scientific, or political stakes in agbiotech – this may be a reasonable description of Italy before the events of 1996–1997. The Italian parliament belatedly approved its first biotech legislation in March 1993 and modelled it closely on the EU Directive 90/220 which had left considerable scope for national interpretations (Commandeur et al. 1996). On the other hand, despite the environmental orientation of the directive, the competent authority in Italy was to be the Ministry of Health, and its advisory committee included many medical specialists – and only one ecologist. If any criticism was voiced about the regulatory framework, it mainly originated in the environmental movement and highlighted the lack of active monitoring and regular inspections of research facilities. Nonetheless, as Commandeur et al. (1996) concluded at the time, '[s]o far, no application for deliberate release has met serious public opposition.'

General public awareness of or interest in biotechnological innovations continued to be very low right up to the 'years of controversy'. Debate was limited to small groups and focused almost entirely on issues of 'red' biotechnology, particularly human reproduction. The natural environment itself, in accordance with traditional Catholic emphasis on human-centred bioethics, did not feature in the early Italian discussions (van Dalen 1997). The arrival of GM soya did not raise many eyebrows in Italy (Grabner et al. 2001), but then the story about 'Dolly the sheep' (the first-ever cloned mammal) broke in February 1997. Inspired by growing public outrage in France and other neighbouring countries, Italians began to question the economic logic behind and the moral justifications for genetic modification of plants and animals (Saba et al. 1998). Consumer and environmental groups, as well as the small farmer association Coldiretti, seized this 'window of opportunity' and formulated a narrative that connected the defence of Italian 'agro-food patrimony' with the moral qualities of the 'natural', diverse, and artisanal (Kurzer and Cooper 2007).

A 'northern' example of how the 'years of controversy' reinvigorated public opposition to GMOs is the case of Denmark. The country's 1986 law on gene technology and environment constituted the world's first dedicated regulatory framework (van Dalen 1997). A relative consensus on this legislation emerged from a wide-ranging national debate with extensive public participation. The generally cautious outlook did not disappear over time, but industry lobbying and rising public confidence allowed for subsequent easing of regulations. However, this period proved to be short-lived and the popular responsiveness of Danish legislation now led to a speedy revision of biotechnology regulations (Grabner et al. 2001). The arrival of 'Monsanto's soy' in Europe undermined Danes' trust in their policy-makers' assurances and the country's ability to remain GM-free in the context of the European Single Market. In the Eurobarometer 1996, public acceptance of agbiotech in Denmark once again ranked among the lowest recorded in the whole of Europe (Bauer et al. 1997). The government's response to the upsurge in scepticism was swift and congruent with the Danish political context. Danish politicians became the public's advocates in Brussels, making the case for broader type of risk assessment that took into account complex ecological systems, ethical objections, and compatibility with goals such as sustainable agriculture. What deserves special emphasis is the fact that Danish resistance was not simply 'created' by the events of 1996–1997 but reflected the persistent strength of underlying attitudes. Zechendorf (1994: 874), for instance, recorded Danish surveys from the late 1980s which show that around 75% of the population disliked interference with 'nature'.

The overall European picture is thus a mixture of convergent and divergent trends. My emphasis on common European trends is not meant to obscure intra-European diversity, which will be discussed later in more detail. Nevertheless, it may be worth going back to broader and long-term quantitative indicators, such as an index of technological optimism between 1991 and 1999. This index reveals that, in the EU, all major technologies (e.g. computers, solar energy, space exploration) were associated with relatively stable measures of optimism, whereas biotechnology fell from 50% to 40% and pessimism rose from 11% to 25% (Gaskell et al. 2001: 55). This observation not only suggests the possibility – at some level of abstraction – of a common European narrative but also disconfirms the view of the watershed years as a mere 'blip' in the public consciousness. Together with the emergence of greater 'traditionalist' resistance in the European north and an incipient opposition to agbiotech in southern Europe, an intra-European cleavage had been substantially weakened (Gaskell et al. 2001), as demonstrated by the 'synchronisation of media coverage across many countries' (Bauer et al. 2001: 50). Two 'distinct plots' now deal with agbiotech and biomedical applications and have largely abandoned the optimistic storylines of uncontroversial progress, even though 'red' biotechnology is frequently presented as beneficial.

At the same time, policy-makers in Brussels struggled to keep pace with the upsurge of popular antagonism. The result of their efforts, the 1997 Novel Foods Regulation (258/97), was outdated even before it came into force, for the member states had begun to disagree on GMO authorisations on a regular basis. The regulation itself did not so much reflect a response to the events of 1996–1997 as a desire to harmonise European approaches to 'novel foods' (including GM foods) without imposing overly stringent regulations. It provided for both specific risk assessment of such food products and their labelling. The important difference from the later Directive 90/220 was that a simple notification procedure (without labelling) was intended to cover those foods that were derived from GMOs but did not contain them in the final product. The Novel Foods Regulation drew on the notion of substantial equivalence of 'existing foods or food ingredients as regards their composition, nutritional value, metabolism, intended use and the level of undesirable substances contained therein' (Article 3). It thus moved the European framework closer to the core assumptions of US agbiotech regulation. Although this simplified procedure would only apply to a minority of products, the intention of boundary-drawing was strongly contested by the European Parliament (EP) and NGOs (Echols 1998). Nevertheless, one may speak of a regulatory 'sea change' in the sense that food safety had previously been based on experience rather than explicit and meticulous scientific risk assessment (Bernauer 2003: 46).

Considerable debate on the new provisions was also evident in negotiations among EU member states themselves. The upsurge in public resistance to GMOs had hit some countries harder than others, and the Commission did not yet feel compelled to exercise restraint: it duly approved varieties of GM maize and GM soy in 1996, despite the opposition or abstention of 14 (out of 15) member states in the Environment Council. A number of member states then began to impose unilateral bans on the approved GM plants. Under the 'safeguard clause' of Directive 90/220 governments were entitled to do this if they had justifiable reasons. The justifications given were, of course, debatable, but there was no desire to force the recalcitrant countries into compliance. And the first signs of compromise embodied in the Novel Foods Regulation proved woefully insufficient in allaying growing antagonism across the whole of Europe (Skogstad 2006: 233). Rather than giving some breathing space to European political elites, it appeared to demonstrate their unwillingness to take full account of citizens' preferences and widespread, multi-faceted concerns over GMOs. As 'a wave of scepticism flowed over Europe, the process of product approval ground swiftly to a halt' (Grabner et al. 2001: 28). Consequently, member state obstructionism continued to strengthen and, at a meeting of the Environment Council in early 1998, eventually spilt over into an informal de facto moratorium on new GMO authorisations. This measure was supported above all by Austria, Denmark, Greece, Italy, France, and Luxembourg and was welcomed by the increasingly pan-European anti-GMO movements. Numerous opinion surveys revealed that the trend towards greater public hostility had by no means reached its peak, at least not until the early 2000s. The Eurobarometer 2000 suggested that GM foods would be rejected by a clear majority of consumers even if they were to offer better taste or lower prices. Keeping in mind large variations across countries, the overall figure of 70.9% for consumer rejection (from Eurobarometer 2001) represented another powerful warning shot for biotechnology promoters (Bonny 2003).

GMO regulation into the new millennium: From directive 2001/18 until 2008

A significant number of European Directorates-General and national governments were still committed to a future for agbiotech and also feared the impact of the moratorium on trade relations with the US. With the Novel Foods regulatory regime 'virtually inoperable' in a climate of increasing distrust (Bernauer 2003: 46), decision-makers offered far-reaching legislative changes. The new Directive 2001/18, replacing the first version (90/220), has been interpreted as a significant break with the historical policy style of the EU. Skogstad (2006: 234) argues that the 'technocratic regulatory style lost legitimacy and precipitated a legislative process and a GM regulatory framework consistent with a mediative regulatory policy style of consensus-building across state and societal actors.' Having failed to declare member state bans illegal by an appropriate majority in the Council, the Commission showed itself flexible and responsive. Recognising labelling as a major bone of contention, it offered Directive 97/35 (on adaptation to technical progress) as a 'sweetener' that would bring in the additional label of 'may contain GMOs' for products merely derived from GM ingredients but not made from them. This measure was taken up by the Council and refashioned into a new Regulation 1139/98 which required the label 'produced from genetically modified organisms'. At the same time, however, efforts were underway to 'de-politicise' food safety regulation in the Union in order to possibly avoid future internal and external regulatory conflicts. In January 2000, the Commission proposed the establishment of EFSA which would take charge of 'risk assessment' while leaving 'risk management' in the hands of elected politicians. In essence, this was a dynamic of trade-offs. Although the Commission was increasingly responding to the wishes of resistant member states and their publics, it was also intent on transforming the structural basis of food safety policy and hoped to re-write the 'rules of the game' in a manner that would allow both further technological progress and popular acceptance.

These changes bought some time to allow for the negotiation and implementation of the new Directive 2001/18 which contained a whole raft of significant amendments: it (1) enshrined the precautionary principle as an overall rationale; (2) provided a detailed eight-page annex with instructions for environmental risk assessment covering long-term impacts, post-release monitoring, and traceability; (3) included a preamble referring to respect for ethical principles and consultation with relevant ethics committees; (4) posited a mandatory consultation of both the public and relevant scientific committees; and (5) imposed further restrictions on the use of antibiotic resistant marker genes in GMOs while stipulating a ten-year time limit to all approvals. The approval process itself still involves two steps: once a competent authority in any member country has submitted an assessment report to the relevant EU committees, discussions are held and, in the absence of agreement, the Commission draws up a proposal. This could now be rejected by the Council with a qualified majority.

By the time the new regulation took effect (October 2002), public antagonism towards GMOs had arguably passed its peak and began a gentle decline. Yet, regulatory wrangling continued. Many governments who staunchly supported the de facto moratorium (i.e. Austria, Germany, France, Italy, Luxembourg, Greece, Denmark) were not willing to admit the revival of authorisations before further regulatory measures were in place. Regulations 1829/2003 (on GM Food and Feed) and 1830/2003 (on Labelling and Traceability) extracted GM foods from the remit of the Novel Foods Regulation and finally broke with the principle of substantial equivalence, thus completing the establishment of an entirely separate legislative category (Tsioumani 2004: 286). The regulations furthermore required the labelling of GM animal feed, so that consumers might choose to avoid meat from animals fed on a GM diet, although there is currently no obligation to label meat products themselves. The threshold for 'adventitious presence' for food and feed was reduced from 1% to 0.9%, and feeding tests on animals and 'in vitro' experiments have become mandatory for the authorisation of any new GM product. A final, potent addition was the new 'contamination' clause, whereby governments are entitled to take protective action – by passing 'coexistence' measures – if they consider conventional and organic agricultural practices to be threatened by GM 'contamination'.

As a result of this further regulatory upgrade, April 2004 witnessed the official end to the de facto moratorium: a pest-resistant GM maize by Syngenta was approved by the Commission. Meanwhile, the Commission continued to pursue a triple strategy of compromise, centralisation, and enforcement. Another attempt was made in June 2005 to order recalcitrant countries to lift their national bans on certain GM varieties within 20 days. The timing was not coincidental, as the WTO ruling on the US complaint against the EU was still undecided at that moment. The outcome of the request was a stinging rebuff to the Commission, with a huge majority of 22 ministers in the Environment Council voting against its proposal.

Although the WTO dispute clearly overshadowed the regulatory debate during these years, the eventual ruling in the autumn of 2006 did not have a tangible effect (see Introduction). The EU responded to the WTO's verdict by arguing that the dispute did not address its updated regulatory framework for agbiotech and insisted that there were no grounds for rethinking its current approach. While the US continues to criticise the EU's regulations, the Union has managed to settle the issue with co-complainants Canada (July 2009) and Argentina (March 2010) by promising to increase the level of consultations and to hold bi-annual meetings among competent authorities.

Although the EU has thus acted as an effective defender of its member states' interests on the international stage, its long-term strategy for agbiotech is decidedly more enabling. The flipside of the regulatory compromise is that the two 2003 regulations equally marked a shift towards greater centralisation. Not only do they apply directly without the need for transposition into national legal codes, but they also imply a tightening of the 'safeguard clause' (which now overtly requires the Commission's tolerance) and a greater role for new, central regulatory body, the EFSA, based on Regulation 178/2002. EFSA's opinions on new approvals now play a key role in risk assessment and are presented to the member states as authoritative, professional knowledge, thus entering into competition with national scientific committees (Pollack and Shaffer 2009). Perhaps because EFSA's members appear to be more suited to assessing pesticides than the complex science of plant genetic engineering (Toke 2004: 169), the agency has almost never produced a negative judgement on a GM crop.

Many member states have been less than enthusiastic about EFSA's arrival on the regulatory scene, which partly explains the agency's limited influence. Distrusting the very idea of a centralised 'oracle', they have been eager to preserve the political-scientific legitimacy of national food authorities (Alemanno 2006). National experts are challenging EFSA's opinions on a regular basis and thus demonstrate ingrained interpretive practices at the centre of scientific risk assessment. Nevertheless, when decision-making in the Council has stalled, the Commission has heavily relied on EFSA's work to justify the approval of GM varieties for import or even cultivation – despite Regulation 1829/2003 also referring to 'other legitimate factors' such as European consumer interests (Pollack and Shaffer 2009).

The age of coexistence? From 2008 towards partial renationalisation

Over the last few years, EFSA has stepped up its cooperation with national food safety authorities and has sought a regular dialogue with environmental NGOs, consumer groups, and industry. But suspicion over systematic bias lingers, and the agency will find it difficult to acquire wider political legitimacy (Klintman and Kronsell 2010) while accusations of 'regulatory capture' by industry lobbies have not been convincingly disproved. Diána Bánáti, the chairperson of EFSA's management board, resigned in 2012 after failing to declare her membership of the industry-funded International Life Sciences Institute. Despite introducing stricter rules on transparency, a report by the Corporate Europe Observatory from October 2013 argued that conflicts of interests were rife, with almost 60% of experts on EFSA's panels having direct or indirect links to the food and feed industry (Euractiv 2013).

Besides the vexed question of risk assessment, by 2008 it had become clear that political tensions were not confined to the approval process for GM crops. Realistically, the total area of GM crop cultivation could only grow in the future if enough consumers were ready to buy GM products and if the supply side could accommodate different agricultural paradigms without leading to ruinous litigation over 'contamination' of non-GM crop varieties or biodiversity (Levidow and Boschert 2008). The central dilemma of the new agenda is that coexistence essentially remains an 'empty signifier' that can be adopted by different interest groups. Whereas the Commission understands coexistence to mean 'harmonious cultivation', the 2010 Cultivation Package grants a very considerable degree of flexibility to member states who are empowered to use a wide variety of measures to ensure the economic viability of different agricultural practices (Dobbs 2011). By 2011, 16 member states had adopted coexistence legislation, ranging from relatively permissive rules in Spain (sowing the seed of conflict between different agricultural systems) to restrictive policies in Austria and even a de facto ban in Italy – largely due to a lack of codified rules on coexistence (Levidow and Boschert 2008; Chiarabolli 2011). Many localities and regions across the EU do not believe in the feasibility of coexistence and have set up GMO-free regions, for instance, by banning the planting of GM seeds (Kurzer and Cooper 2007).[1]

While justified in terms of diverse environmental conditions and the management of socio-economic consequences of inadvertent 'contamination', the 2010 Coexistence Recommendation also constitutes a step towards subsidiarity and decentralisation in favour of national governments and, to some extent, sub-national authorities. Dobbs (2011) interprets this as a longer term strategy for re-harmonisation, not least because coexistence may eventually lead to either large-scale GMO 'contamination' or the spread of GMO-free regions. But the spirit of compromise shown by the Commission may indeed be the only short-term solution to the continuing political gridlock at the EU level. Frequent recourse to the 'safety clause' by member states and the tendency to blame the Commission for new GMO approvals (when the Council is deadlocked) made it clear that the status quo was a poor example of European regulatory harmonisation.

Taking up a French initiative from 2008, the Commission published a comprehensive 'Cultivation Package' in March 2010 which entailed a significant renationalisation of risk management and regulation of GMOs and signalled a willingness to accept other than scientific grounds for restricting the cultivation of GM crops, such as socio-economic or cultural reasons. The package was welcomed by a small number of countries intending to either ban GMOs outright or promote them. But it was equally criticised by many supporters as well as opponents of GMOs, with concerns including the preservation of a centralised authorisation process (with the Commission and EFSA as key players), the vagueness of permissible grounds for restricting GM cultivation, and the potential threat of WTO trade sanctions against individual European countries. Although subsequently 'fortified' by the Environment Council and the EP – for example with provisions for explicitly environmental objectives – a Danish-led compromise in March 2012 came to naught. After a period of uncertainty, renewed interest was sparked by the ruling of the ECJ on Bt-1507 maize, forcing the Commission to table an authorisation for cultivation of a crop that had languished in the approval system since 2001. With Germany declaring its support for regulatory reform, the Greek Presidency published a new compromise paper in March 2014 which was agreed in June 2014. However, because the European Parliament will also be involved, it remains unclear at the time of writing when (and in what form) a partial renationalisation of GMO regulation will take shape.

Conclusion

The EU's regulatory framework on agbiotech thus remains in relative flux, although it is difficult to imagine a return to the permissive ambitions of the early 1990s. Bureaucratic politics and regulatory outcomes in the US and the EU have often differed quite sharply, as shown by the EU's 1990 regulatory framework under the guidance of DG Environment and by the White House OSTP's technology-friendly guidelines. However, if trade and industrial policy officials had their way on both sides of the Atlantic, they would likely agree on steps towards greater transatlantic regulatory coordination (Murphy and Levidow 2006). The strongest markers of difference are therefore public opinion, NGO mobilisation, political pressure, and consumer activism.

In the US, public opinion surveys yield somewhat ambiguous results but can be interpreted as offering a broadly accommodating political context – at least in the absence of major regulatory failures. Opposition from most citizens appears to be of relatively low intensity and largely focused on physical risks, while the potential benefits of agbiotech tend to sway a significant portion of sceptical respondents. Legal action, commercial priorities, and concerns over key export markets have so far influenced the trajectory of agbiotech more strongly than public opinion, but political mobilisation has considerably strengthened over the last few years. By contrast, in Europe, there was latent public opposition as early as the 1970s. Meanwhile, a technocratic style of regulatory governance, steeped in environmental precaution, prevailed at the EU level, while various political compromises were negotiated at the national level. Yet, everything changed in the mid-1990s. During the 'years of controversy' (1996–1997), latent public opinion was mobilised by NGOs who highlighted the physical risks and, crucially, the moral and cultural implications of the agbiotech revolution. A new regulatory framework adopted in 2001 demonstrated the limits of industrial policy objectives and WTO pressure and led agbiotech proponents to pursue coexistence with other agricultural systems. By the 2010s, the uneasy political compromise at the heart of the EU's regulatory framework had run out of steam, and there seems to be increasing momentum behind plans for a partial renationalisation of GMO regulation. Overall, however, European public opinion has remained sceptical of agbiotech. The Eurobarometer 2010 indicated that 27% of respondents voiced support and 57% registered their disapproval (Gaskell et al. 2010).

2
Perspectives on Regulatory Divergence

Introduction

The observation that public attitudes towards agbiotech are relatively well correlated with regulatory outcomes does not yet establish the primary relevance of public opinion. Hence, this chapter seeks to review and learn from existing explanations of transatlantic regulatory divergence. There are many strands of regulatory studies, but scholars exploring the politics of agbiotech often employ pluralist and institutionalist perspectives on political science. The former are dominated by political-economic analyses. Within the latter camp, three variants of 'new' institutionalism – rational-choice, sociological, and historical – are of particular relevance for the study of agbiotech policy-making.

Both institutionalist and political-economic accounts significantly enhance our understanding of regulatory dynamics in the US and the EU. Given the political salience of GMOs in many countries, societal preferences have clearly had an impact on the regulatory process – often through sustained public mobilisation in the shape of consumer boycotts and political lobbying. Political economy perspectives are particularly good at linking societal and interest-group pressures to political outcomes, but they are not especially concerned with exploring the nature and intensity of public attitudes towards GMOs (except Schurman and Munro 2010). In the EU, public anxiety caused by food crises and media 'scare stories' – with a resulting loss of trust in scientists and regulators – have played an important part. But these aspects do not seem to be decisive on their own. They only matter alongside a top-down, morally infused process of attitude formation which provides an interpretive 'filter' for GMO-related events and which has often led to 'moral vetoes' among Europeans. In the US, moral concerns have been more effectively counterbalanced by considerations of 'usefulness' (such as practical benefits, price reductions, etc.) – even in the face of GMO-related 'events' with crisis potential. After beginning with a brief overview of institutionalist approaches, this chapter explores political-economic perspectives and finally delves into a detailed analysis of public attitudes towards GMOs, covering conventional risk–benefit evaluation, trust-related perspectives, and higher order ethical/moral considerations.

Institutionalist perspectives

Rationalist institutionalism

Beginning with the broad category of rational-choice institutionalism, the comparative analysis of political systems draws attention to the centrality of administrative and regulatory arrangements as well as the resultant political opportunity structures for actors. Federal systems are commonly portrayed as offering multiple venues for influence and opposition which might lead to policy deadlock in the short term and political compromises in the longer term – and thus to more stringent agbiotech regulations. But the nature and effects of 'regulatory federalism' (Bernauer 2003) are not analogous on both sides of the Atlantic. While both political systems are models of 'fragmented governance' (Varone et al. 2007), the US has not witnessed high levels of internal regulatory competition. Bernauer (2003: 12) argues that the US regulatory system is characterised by 'centralised laxity', as most state-level legislative initiatives have served to further protect GM crops (e.g. against vandalism) and because more stringent legislation would likely be challenged in court.[1] Sheingate (2009) notes that the administrative foundations of US agbiotech policy are based on a patchwork of regulatory authorities which has the effect of diminishing both the extent of state involvement and the authorities' financial resources. Nonetheless, the regulatory philosophy of the agency or ministry in charge of drafting and implementing agbiotech regulation will always be an influential factor (Varone et al. 2007).

Conversely, the EU's quasi-federal political system has led to a 'ratcheting-up' of agbiotech regulations. Notwithstanding the prominent role of the Commission with its privilege for legislative initiatives and a hitherto muted role for the ECJ, the EU's member states are also heavily involved in policy-making. They are both the principal agents of enforcement and assume formal political roles in European regulatory committees and, ultimately, the Council of Ministers. A critical mass of sceptical member states was instrumental in supporting a precautionary regulatory attitude (Tiberghien 2009). Moreover, European citizens are represented by the European Parliament which treasures its longstanding image of being the advocate of their concerns and operates under the co-decision procedure when it comes to regulatory reform (but not regarding the approval process for GM crops). In sum, the type of political system certainly matters for regulatory policy-making, and particular attention must be paid to the precise distribution of authority and resources among the different actors and at different levels of governance. However, as a sole variable, political systems do not have a strong predictive quality. Examining different national agbiotech regulations (France, Sweden, the Netherlands), and bearing in mind the limited scope for diversity under EU directives, Montpetit et al. (2007) find that more centralised (and less fragmented) governance does not necessarily imply more permissive regulations.

Sociological institutionalism

Some of the structural, contextual forces that work alongside administrative and representative frameworks are captured by sociological institutionalism – a theory that focuses on 'the forms and procedures of organizational life stemming from specific cultural practices' (Schmidt 2010: 13). The term 'cultural' is understood here as referring to a 'logic of appropriateness' (March and Olsen 1984) rather than to the rationalist logic of utility maximisation.

Scholars have identified divergent 'policy styles' which denote 'central and distinguishing features of policymaking and implementation' in the US and the EU (Skogstad 2006: 215). These styles are also heavily shaped by a polity's historically evolved practice of food safety regulation and its dominant understanding of the role of science and technology. Skogstad thus identifies a US policy style marked by 'adversarial legalism' whereby a host of dispersed and often underfunded state actors seek expertise and support from other bureaucratic and societal actors. The constant threat of litigation against inconsistent regulations or regulatory failures and food scandals exerts considerable discipline on both public and private actors. In the 1980s, when the Reagan Administration was prepared to install an extremely permissive regulatory system, it was private corporations such as Monsanto – undoubtedly concerned about the risk of litigation – which insisted on specific, legally defensible rules and regulations (Schurman and Munro 2010: 45). In the EU, Skogstad observes a consensual, mediative regulatory policy style which is conditioned by dispersed political authority and the need for unanimity or supermajorities to pass legislation – as well as on member state cooperation to implement rules and directives. A mediative policy style is equally required due to the diversity of national goals and values, the brittle legitimacy of harmonised European regulations, and the resulting tendency to avoid sharp conflicts over policy.

In a more substantive sense, one can equally think of policy styles as embedded within broader 'regulatory cultures'. Regulatory frameworks may be understood as resting on 'major value orientations', for instance expressed in terms of different levels of tolerance for uncertainty and risk (Wohlers 2010). It has often been argued that the US regulatory system relies on a paradigm of 'sound science', whereas the EU pursues a more precautionary approach characterised by relatively permeable boundaries between natural science, social concerns, and policy outcomes (Jasanoff 2000; Ramjoué 2007). In the US, risk tends to be defined more narrowly as the probability of harm to humans or the environment – a conception which excludes broader social considerations of risks/costs and benefits. In the mid-1990s, the White House explicitly stipulated that federal regulatory agencies were not permitted to take into account the socio-economic impact of their decisions (Gaskell et al. 2002: 372).

There are evident parallels of sociological institutionalism with the cultural analysis of agbiotech policy pursued in this book. This becomes especially apparent when scholars move from the meso-foundations of regulatory culture to more encompassing notions of 'political culture' and 'national identity' (Jasanoff 2005; Wohlers 2010). I will lay out my own take on cultural politics in Chapter 3.

Historical institutionalism

A third strand of institutionalism seeks to integrate elements from both rational-choice and sociological perspectives, while also accounting for the resilience of regulatory frameworks and the potential causes of regulatory shifts. Historical institutionalism conceptualises the evolution of regulatory governance as a series of punctuated equilibria, with significant shifts only occurring when 'policy windows' open up – often as a result of an 'exogenous shock' such as a regulatory failure in food safety policy (Greener 2005). At the heart of the theory is the contention that initial decisions – notably the adoption of a particular regulatory framework – create a path-dependent logic of regulatory evolution. Once a framework with a distinct regulatory philosophy has been instituted and is overseen by a set of bureaucratic actors, significant shifts will not only be difficult, but any potential changes will also be shaped by the prevailing regulatory patterns. The moments when these regulatory paths are set or shifted are called 'critical junctures'. In the case of agbiotech, critical junctures can be located in the 1980s for the US and a few years later in the EU when the foundational regulatory frameworks were put in place. The second critical juncture for the EU came during the second half of the 1990s amid the 'years of controversy' when agbiotech supporters were rebuffed and the framework was made even more stringent (Pollack and Shaffer 2009: 12–13).

Once regulations are in place or reforms have become embedded, historical institutionalists trace the means by which prevailing regulations become 'locked in' or self-reinforcing. It is assumed that policy reversal quickly becomes too costly because bureaucratic actors, interest groups, and the wider public adapt to and become invested in the new regulatory environment (Sheingate 2006; Pollack and Shaffer 2009: 79). The agbiotech industry in the US would thus pour money into research and commercialisation, while actors across the EU food chain would hesitate or focus on conventional and organic agriculture. As for the lead regulatory agencies, USDA and FDA would be expected to defend their position and permissive regulatory philosophy, and the same applies to the EU's more precautionary DG Environment.

However, these 'positive feedbacks' and internal adaptations are not the only way to explain regulatory stability and 'lock-in' over time. Both policy stability and infrequent regulatory shifts are powerfully shaped by contingent events which disrupt the routine patterns of the agbiotech policy subsector. Burns (2012) cites EU enlargements in 1995 and 2004 – as well as the introduction of co-decision powers for the European Parliament – as catalysts for a precautionary orientation. Furthermore, the perceived food crises in the 1990s over dioxin contamination, mad cow disease (BSE), and the import of Monsanto's GM maize qualify as exogenous shocks which stabilised the precautionary policy framework and made it even more stringent. If significant regulatory change is to occur in the US or the EU, it may well proceed incrementally through a shifting balance of political power, technological or scientific discoveries (including new risks from agbiotech; Pollack 1996). The stronger the policy lock-in, the greater the likelihood that such change will depend on compelling 'external' events which have the necessary force to destabilise established routines and adaptations. The EU's regulatory history offers some examples, but in the US, none of the potential regulatory failures to date – such as contamination of food or feed with unauthorised GM varieties – has led to deeper perturbations.

Overall, institutional analyses of agbiotech policy-making deliver valuable insights and explanations. Yet, frequently they do not sufficiently examine broader state–society interactions which are central to relative policy stability in the US and a precautionary trend in the EU. While rational-choice and sociological institutionalists are mindful of the limited explanatory power of political systems or regulatory cultures, historical institutionalism tends towards more encompassing explanations of agbiotech policies. Thus, Sheingate (2006: 245) suggests that US agbiotech policy – including interest group behaviour and public opinion – 'may be an effect of past policies reinforced by the institutional features of the administrative and congressional policy process.' Similarly, Burns (2012) argues that, in the EU, relatively small decisions at the meso-level which established a policy trajectory in the late 1980s (with DG Environment in charge) were as important as the larger, exogenous shocks which occurred later on.

Regarding Burns' first point, it is justifiable to highlight crucial meso-level political decisions. Consequently, one also needs to take into account the impact of entrepreneurial political actors. Daviter (2009) draws on bureaucratic politics and discourse theory to argue that the allocation of responsibility within the European Commission was crucial for determining which political narrative – economic competitiveness or risk management and public health – came to dominate. The power of problem definition and framing allowed some DGs within the Commission to change the perceptions of other bureaucratic and societal actors. Eventually, in the late 1990s, the European Parliament's ethical framing of biotechnology gathered strength, and DG Environment and DG Health and Consumers SANCO prevailed by pointing to rising public concern and insisting that agbiotech could only be successful once consumers felt sufficiently protected by a revamped, more stringent regulatory framework.

However, it is likely that the success or failure of framing strategies was very closely linked to other political forces – such as national governments, the industry lobby, NGO campaigns, and consumer boycotts. One way of further expanding the scope of analysis is to explore the influence of 'policy networks', conceptualised as 'clusters of public and private actors connected to each other by resource dependencies, such as information, expertise, money and legitimacy' (Carter 2007: 187). Policy networks operate in many European countries as well as the US, and agricultural policy-making, in particular, has often been heavily shaped by them. With many different actors involved, including the supranational food safety agency EFSA (since 2002), there is no genuine policy network to speak of at the EU level. But there certainly is a relatively cohesive and influential group in the US – the 'Food Chain Network' (Toke 2004: 133) – and there has also been a policy network on agbiotech in the UK. This particular network underwent significant change in the late 1990s by widening its remit and including several nature conservation organisations in the formal regulatory committees. While an exploration of policy networks may enhance our understanding of micro- and meso-level political dynamics – and will briefly feature again in Chapter 4 – institutionalist and policy (network) analyses typically focus on the impact of bureaucratic actors and other 'insider' groups. This should be complemented by more sustained analysis of various 'outsider' groups, such as more radical environmental NGOs, and of broad-based public mobilisation (Toke and Marsh 2003).

As for Burns' (2012) second point about critical junctures and exogenous shocks, it is important to recognise that the identification of critical junctures and the evaluation of their long-term effects remains an interpretive exercise. Building on sociological institutionalism's interest in culture, it is uncontroversial to say that initial regulatory frameworks were not constructed in a cultural and political vacuum. In other words, the path dependence theorised by historical institutionalism could be seen as part of an approach based on cultural and political contextualism. As some institutionalists acknowledge, rather than focusing on a particular administrative decision, scholars also need to explore the *context* of early policy choices – institutional and normative (Skogstad and Moore 2004: 49), as well as historical and socio-cultural (Sheingate 2009: 493). And, although contextual shifts are possible and may enable policy changes, contextual conditions are best conceptualised as structural conditions that reach much further back into history (see Chapter 3). Likewise, the insistence on exogenous shocks needs to be qualified because particular events have to be interpreted as regulatory failures by the media, policy-makers, and the wider public before they can exert their transformative or stabilising effect. Europeans were receptive to such shocks, as they had already developed a pervasive sense of a wider 'food crisis' by the 1980s, particularly in relation to mass-produced, 'industrial' food (Toke 2004: 99). Overall, therefore, it is best to follow Pollack and Shaffer's (2009: 76) argument that contingent events did not directly cause US or EU policies, but provided new opportunities and resources for political actors competing over problem definitions and policy measures. A broader evaluation of other actors and the impact of consumer power should therefore encompass political-economic perspectives as well.

Political economy

Political economy perspectives provide both actor- and society-centred explanations of regulatory dynamics, directing our attention towards business lobbies, farmers' organisations, and NGO coalitions. Some accounts emphasise the role played by powerful corporations and farmers' organisations in pushing for the adoption of permissive regulation in the US and stringent regulation in the EU. This assumption is largely correct for the US, where the agbiotech-friendly 'Food Chain Network' – encompassing input suppliers, food processors, retailers and farmers – has strongly influenced policy-making at federal and state levels, while also spending considerable amounts of money on generating public support and countering anti-GMO activism (Bernauer 2003: 94–95; Toke 2004: 133). The largest farmers' organisations in the US have typically thrown their weight behind GM seeds and the promises for higher yields and lower costs (of using fewer pesticides), although the declining availability of conventional seed varieties may have equally been a factor.

Organised interests: Biotech industry and farmers' organisations

However, there is little evidence for the claim that a similar alliance of organised interests has been instrumental in bringing about the EU's shift towards stricter regulation. Graff et al. (2009) argue that rational and converging interests of several actor coalitions ultimately explain the transatlantic regulatory divide. Lagging behind the US in agbiotech patents and innovation capacity, European companies continue to rely on their comparative advantage in chemistry (and conventional pest control products). Farmers, meanwhile, would be protected from potentially cheaper imports of GM crops, and NGOs could enhance their funding base by spearheading the anti-GMO movement. This reading of European developments is intrinsically coherent, but a closer look at the industry's lobbying activities yields a different assessment.

First, the biotech industry's long-term economic strategy was not advanced by strict regulations. European firms might gain protectionist benefits from restrictions on GMO imports, allowing them to either sell more non-GM products or build up their own research and development (R&D) capacity. This scenario – together with concerns over access to export markets – may partly explain China's turn towards more restrictive regulations in the early 2000s (Macilwain 2003). However, in a highly internationalised and competitive sector, the temporary halt to GM crop authorisations was a costly setback for European companies such as Syngenta or Bayer (Rosendal 2005; Falkner 2009). At the global level, biotech corporations have usually banded together to defend the interests of a globalised industry that needs stable markets to recoup costs and re-establish the degree of profitability that conventional pesticide patents have ceased to provide (Andrée 2007). By moving some of its R&D facilities to the US, Europe's biotech industry demonstrated that it was not prepared to exit the sector and forego the future potential of agbiotech products.

Second, lobbying efforts have not been intentionally ineffective. Overestimating the relative power of DG Research and DG Enterprise within the European Commission, Europe's biotech industry belatedly created its own lobbying organisation, the Senior Advisory Group on Biotechnology (SAGB), in 1989.[2] Its later incarnation – EuropaBio from 1997 onwards – is significantly more unified internally and, by 2013, counted 55 corporate members and 17 national biotechnology associations. This is especially important because divisions between smaller and larger firms as well as between agricultural and pharmaceutical biotech sectors had long dogged European lobbyists. Overall, the industry's readiness for compromise was a product of the challenging circumstances. It 'persistently lobbied for laxer approval and labeling regulations' and only grudgingly accepted mandatory labelling (Bernauer 2003: 83). By the turn of the century, members of EuropaBio had become so impatient about the de facto moratorium on new GM crop approvals that their main focus was on resolving the political gridlock (Rosendal 2005). The revised regulatory framework was much closer to NGO demands than industry preferences, and it can therefore be seen as an ill-fated attempt at damage limitation.

Third, it is true that European farming interests often deserve their protectionist reputation and that, in contrast to the US, the EU's agricultural system has been gradually moving towards less intensive practices (Morgan et al. 2006: 27; Ramjoué 2007). Generally, however, farmers' organisations have not been the protectionist bedrock that some had expected. Many small farmer NGOs and organic advocacy groups have been vociferous in their criticism of agbiotech, but mainstream representatives have at most displayed a mild scepticism about the grand promises of the agbiotech industry. Their support for labelling and liability laws is rather a response to consumer preferences and the need for a non-divisive policy of coexistence (between GM and non-GM crops) than a genuine commitment to a GM-free Europe (Ansell et al. 2006: 101). In fact, European farmers may even be forgoing a much-needed boost of their competitiveness by enduring the strict regulatory regime and agreeing to liability rules (Bernauer 2003: 84).

Relative openness to GM technology has been corroborated in a study by the University of Göttingen, Germany, which found that 47% of British farmers 'generally favour GM crops' and that almost three-quarters of them would grow them 'if consumers were willing to buy them' (GMO-Compass 2007). In Germany, the majority of farmers were still undecided in 2007. More recently, 59% of Greek farmers, almost half of Czech farmers, and one-third of farmers in Spain, France, and Hungary expressed interest in growing particular varieties of GM crops (Areal et al. 2011; Skevas et al. 2012). Economic motives were predominant, but farmers also frequently believed that GMOs posed risks to public health and the environment. Ultimately, it is reasonable to deduce that the consent of most mainstream farmers to a GMO-averse agricultural policy is conditional upon the absence of lucrative markets for GM produce (Bernauer 2003: 85). Over the last few years, European livestock farmers have joined the chorus of GM supporters demanding speedier approvals of imported GM varieties as animal feed.

In sum, in the European case there is no firm evidence for an explanation based on rational economic interests – protectionist or longer term and strategic. Nevertheless, the underlying intuition remains valid, insofar as agricultural markets can also be subject to a kind of 'path dependence'. A gradual process of adaptation pertains to all actors in the commodity chain. Even if European companies and farmers have initially had little interest in blocking GM technology or reducing GM crop imports (which might turn out to be cheaper than home-grown conventional crops), a strict regulatory system coupled with consumer aversion could lead them to adjust their interests over time. If EU companies and agricultural producers were to fall behind too far and find it too costly to switch to the newer, potentially more efficient, agricultural technologies, then they might begin to mount a robust defence of incumbent, 'old' technologies, such as conventional plant protection methods and organic practices (Swinnen and Vandemoortele 2010). There is scant evidence to believe that this process of adaptation has reached a critical stage, but if non-GM conventional alternatives continue to make significant progress, then sticking with the 'old' technology may remain a viable alternative. However, overall, the strongest political-economic arguments with immediate relevance point to the centrality of 'countervailing forces' and public opinion.

Countervailing forces

The nature and intensity of public opinion, briefly chronicled in Chapter 1, is at the heart of any successful anti-GMO movement with lasting impact on regulatory decisions. Chapter 4 will expand on the movements' strategies which are, to a significant extent, reflections of the societies in which the movements are rooted (Schweiger 2001). This is not only because they are constituted by citizens of particular societies, but also because their success or failure depends on their choice of objectives, tactics, and discourses. Despite a five-year period (1998–2003) which saw intense activity thanks to large inflows of grants for anti-GMO campaigning, US-based NGOs failed to achieve a turning point in agbiotech regulation (Schurman and Munro 2010). Public awareness and concern peaked during that time and discourses about potential environmental and health risks – as well as the need for mandatory labelling of GM products – appeared to resonate with sections of the public. The limited degree of mobilisation, however, was not able to overcome the entrenched regulatory system and its powerful supporters from across the food chain. Even if some consumers were inclined to change their purchasing behaviour, the absence of mandatory labelling made it difficult to apply selective pressure on particular manufacturers. Downstream actors in the commodity chain are relatively dispersed, and the retail sector as a whole is thus not overly vulnerable to competitive pressure from consumers. At the same time, upstream actors are more concentrated and largely unwilling to systematically segregate GM and non-GM varieties (Bernauer 2003: 87–88).

In the EU, by the mid-1990s public opinion did not only directly influence policy-makers and effectively set national and supranational agendas (Prakash and Kollman 2003; Tsioumani 2004). It was also 'activated' in the economic sphere through broad-based, well-crafted NGO campaigns which often targeted food retailers through boycotts. Once labelling had been agreed, a 'domino effect' began rippling through the entire food chain (Rosendal 2005). With market concentration downstream in the retail sector around 60–80% in some European countries, and a comparably low concentration upstream in farm, seed, and grain-handling sectors, there were both strong pressures and realistic market opportunities for sourcing sufficient quantities of non-GM products (Bernauer 2003: 87–88).

As noted earlier, although institutional and economic structures clearly played an important role in enabling or restraining anti-GMO movements, public opinion provided the necessary 'fuel' for political-economic developments to unfold. Admittedly, not all political decisions in the US and individual European countries directly correlate with public opinion data – for instance, at times ministers in Italy and

France seemed to go beyond what was being demanded by the majority of respondents (Hampel et al. 2006). But longer term trends confirm that the scale and intensity of public concerns broadly correspond to regulatory change or stability. Ultimately, more revealing than statistics as crude indicators are the reasons behind the public's disquiet, and these can only be identified by combining survey methods with qualitative data and interpretive approaches. In this vein, several hypotheses have been proposed to explain transatlantic divergence regarding the intensity and persistence of public disquiet.

Exploring the roots of public opinion

If GMOs represent considerable potential for 'public outrage', which appears to be the main driver of European regulatory politics, it is necessary to investigate the causes of this strong and sustained public reaction. In the last section of this chapter, I argue that Europeans display a predominantly moral understanding of agbiotech which, as the remainder of this book will show, is itself pervaded by historically transmitted cultural meanings. Before that, however, a number of alternative approaches still need to be reviewed.

First of all, a conventional way of managing the potential risks of agbiotech would be to assess each of its applications on its own merits. This typically involves a risk–benefit analysis that combines an estimate of the likely benefits with an appraisal of the risks to human health and the environment. Divergent criteria will be applied by different regulatory authorities, but the principles of dispassionate, technocratic risk assessment can be generalised to some extent. However, this conventional approach often does not correspond well to the public's evaluation of agbiotech. There may be a reasonable degree of congruence in the US, but when interpreting the Eurobarometer survey of 1996, Bauer et al. (1997: 845) observe a 'striking mismatch between the traditional concern of regulators with issue of risk and safety, and that of the public, which centres on questions of moral acceptability'. Lassen et al. (2002: 264) similarly perceive an inability of decision-makers, scientists, and the biotech industry to comprehend and accept the ethical character of public opposition.

Scientific literacy and the quest for a 'rational' public

Frequently, such technocratic perplexity is expressed through the 'scientific literacy' thesis. Its proponents are wont to argue that scientific knowledge directly increases acceptance and that the public suffers from a critical knowledge deficit and is therefore unable to make rational decisions about agbiotech applications. One well-developed strand of the scientific literacy thesis examines the impact of education. However, statistical results are generally inconsistent and do not strongly support a focus on education as a central variable. For instance, Puduri et al. (2005) found that, in the US, higher formal education increases approval of plant-based GM products, while Ganière et al. (2006) and Priest (2000) could not establish a significant link of education level with acceptance of GMOs.

The second strand of analysis concentrates on respondents' possession of factually correct scientific knowledge. Comparisons between different statistical results are complicated by the diversity of questions used to measure the level of knowledge. Nevertheless, regarding US public opinion, studies by Puduri et al. (2005) and Ganière et al. (2006) confirmed a positive influence of knowledge on approval ratings. Priest (2000), on the other hand, could only identify a very weak connection. As for the European context, drawing on Eurobarometer surveys, both Gaskell et al. (2001) and Ceccoli and Hixon (2012) found a statistically significant correlation between scientific knowledge and acceptance of agbiotech. By contrast, Pardo and Calvo (2006) only identified a very weak link and, in a recent postal survey in Switzerland, Connor and Siegrist (2010) even found a small negative correlation.

The scientific literacy thesis is closely linked to the 'deficit model' of the public in science communication. Commentators in this tradition are often committed to a linear diffusion model of technological progress, which predicts gradual but steady societal uptake of beneficial new technologies. The 'deficit model' apportions the blame for public reticence both to people's lack of knowledge/awareness and (less frequently) to their lack of trust in regulators (Gaskell and Bauer 2001). The public is cast into the role of the unsuspecting victim manipulated by a host of self-interested powers such as civil society or media organisations. Supporters of the 'deficit model' use arguments about 'emotional' rather than 'rational' reactions drawn from psychometric studies. According to Miller (1998: 14), unknowledgeable consumers tend to rely on a skewed assessment of the risks by neglecting the benefits, overestimating 'invisible threats' and underestimating 'familiar risks'.

But there are justified doubts over the alleged centrality of factual scientific knowledge. First, the diffusion model of technological progress, which the scientific literacy thesis is taking for granted, is often associated with natural scientists 'caught in a social science time warp' – clinging to an oft-refuted theory from the 1970s (Bernauer

2003: vi). Second, the evidence for a correlation between knowledge and acceptance of GMOs is moderately suggestive but not clear-cut, especially for European citizens. Earlier comparative results, backed up by the recent study from Switzerland cited above, showed that Europeans had on average greater scientific knowledge about agbiotech than Americans, yet they were less positive about it (Hallman 2000). In another recent postal survey conducted in Denmark, more scientific knowledge had rather complex effects. It made respondents less likely to distinguish between different kinds of genetic modification (transgenics vs. intragenics), but it also made them more likely to insist on significant societal benefits as a condition for using the technology (Mielby et al. 2013). Third, it is typically assumed that people's 'cognitive deficit' can be remedied by supplying 'correct' science-based information. The US-based International Food Information Council (IFIC 2006) demonstrated that higher awareness is positively associated with purchase intent and that provision of information equally has a positive effect, although the information IFIC provided to respondents was clearly neither 'neutral' nor 'balanced'. Even the provision of less partial information can sometimes make citizens more sceptical, as was demonstrated during the 2003 UK government-sponsored 'GM Nation' public debates. Many of those who had started as undecided participants hardened their views about GM foods (Coghlan 2003).

Media influence

Another frequently used argument is that the public's toleration or rejection of expert-led risk–benefit analysis is driven by sensationalist reporting in the mass media and by tabloid newspapers in particular. One approach focuses on the intensity of media coverage and posits that heightened exposure to media reporting is of particular significance for the formation of public opinion. Early evidence from Europe's 'years of controversy' offers some support for this hypothesis and also appears to show that increased press coverage per se – regardless of tendency/valence – tends to lead to more negative public perceptions (Gaskell et al. 2002). However, this linkage could be confirmed neither by Gutteling's (2005) study of readers and non-readers of popular newspapers, nor by Ceccoli and Hixon's (2012) regression analysis of Eurobarometer data. This uncertainty and the relatively short duration of media peaks about agbiotech mean that there is no unambiguous causality between media coverage and public opinion (Listerman 2010). Even a basic awareness-raising function of the media is not always as effective as one might expect – at least in the US. Temporary peak coverage in the wake of critical events can often overcome the public's apathy and put a topic on its 'radar screen' (Thomson and Dininni 2005: 247). Yet, as Hallman (2000: 2) commented during the height of the earlier anti-GMO campaign in the US, '[d]espite very substantial media attention [...] much of the public seems remarkably uninformed about biotechnology.'

A second approach goes beyond quantitative intensity to explore the content and framing of media reporting. Using a model of 'mediated issue development', Nisbet and Huge (2006) argue that frames – understood not as policy positions but as 'thought organisers' – help to explain why events linked to agbiotech, such as the StarLink controversy of 2000, have not genuinely shaped public opinion. They note that the mainstream media have consistently portrayed GMOs through fairly dry technical and regulatory frames, thus ensuring that the topic would never achieve 'celebrity status'. To increase the topic's resonance, they suggest that agbiotech would have to be linked to other salient issues, for instance to controversial food policy themes. Other scholars have gone even further by employing media agenda-setting theory. They posit that what is reported, and how it is reported/framed, effectively sets the agenda for individuals and policy-makers (Thomson and Dininni 2005; Priest 2006). Examining the US peak coverage of the story on GM pollen and Monarch butterflies as well as on the rejection of US food aid (containing GM corn) by some African countries, Thomson and Dininni (2005) found that US citizens could remember the 'world hunger event' much better and often associated it with important benefits of agbiotech. Tackling starvation in the Third World had much more relevance to their ethical worldview than the Monarch butterfly story of 'ecological loss'.

While the impact of media framing may indeed be substantial in particular cases, there does not seem to be a consistent pattern across different studies. Studies by Gaskell, Einsiedel et al. (2001) and Gutteling (2005) found no support for a strong, uniform, and direct linkage of media content and audience attitudes. Peters and Sawicka (2007: 88) point out that even stories with similar bias (or valence) can, in fact, generate very different evaluative responses by an audience. They assume that readers manage to read between the lines and pick up 'subtle references to existing beliefs, values, and experiences' which then, in turn, generate an overarching meaning for agbiotech.

Although the agenda-setting and mobilising power of the media can sometimes be considerable, in the case of agbiotech the literature does not support a view of the media as chief 'opinion-shaper'. Even if citizens' opinions are often not fully formed and can be influenced, the public is not merely a 'malleable victim' of a sensationalist media. UK focus groups revealed that citizens themselves frequently complained about media sensationalism (Marris et al. 2001: 86). Rather than conceptualising the media–public relationship as a one-way street, it is at least bi-directional (Peters and Sawicka 2007). In political-economic terms, the media itself is heavily shaped by competition and consumer choice, which makes it likely that at least a few media outlets will pick up a particularly salient story about agbiotech – and that this may lead to a 'chain reaction' which quickly ripples through the broader media landscape (McCluskey and Swinnen 2011). In other words, the media's function has more often been that of a mirror of public sentiment – a 'cultural indicator' (Gaskell and Bauer 2001: 6) – than that of a potent politico-cultural actor.

Social trust

Social trust is a third dimension of public opinion which is associated with attitudes towards GMOs. Scholars have often discussed social trust in the context of the 'risk society' thesis,[3] proposing that, in the complex technological environment of modern societies, social trust provides a substitute for knowledge (Gaskell et al. 2002). Consequently, trust could function as an element to further extend the 'deficit model' of the public. In conjunction with exogenous shocks, such as the mad cow crisis (BSE) or the StarLink episode, it is possible to argue that a sudden, substantial loss of trust has made a lasting negative impression on citizens and has delegitimised regulatory authorities for many years to come.

As I set out earlier, however, StarLink failed to provoke sustained outrage, and the event-based storyline about European reactions to BSE does not adequately account for the pre-existing trend towards public disquiet. Rather, UK focus groups portrayed BSE as a regrettable apogee of a long-running trend in food policy (Marris et al. 2001: 84). Nor does the event-based storyline tally with the value-based character of public opposition, as will become clear in the remainder of this chapter. In line with risk society theory, US and European citizens are embroiled in a daily struggle of 'taming the information tide' (Priest et al. 2003: 766) and are looking for strategies that allow them to make the right decisions. Under these circumstances, an exact understanding of the mechanics of biotechnology becomes less important than the belief that the creators and regulators of biotech applications share the same values (Hallman 2000: 5).

The variable of trust is often defined as *institutional* trust – the confidence expressed by citizens in their political and regulatory institutions.

Gaskell et al. (2002) thus reported that US regulatory agencies such as the USDA and the FDA were supported by 90% and 84% of respondents, respectively. The US regulatory system as a whole was trusted much more strongly than, for example, its German equivalent (Peters et al. 2007). The US result confirms a high degree of confidence in the regulators' handling of agbiotech issues. On the other hand, once agbiotech has been strongly politicised and is subject to ongoing debate – that is, in most European countries – institutional trust loses much of its correlation with public opinion (Peters et al. 2007).

As a result, the picture is more complex when it comes to issue- and actor-specific categories of trust which are often summarised as measures of *social* trust. With respect to the US population, Puduri et al. (2005) recalculated survey results of actor-specific trust with the help of quantified attitudes towards biotechnology, estimating that respondents with greater trust in government were 29% more likely to endorse agbiotech, while trust in regulators increased acceptance by 28% and trust in scientific organisations by 66%. Conversely, Priest et al. (2003) sought to directly measure respondents' confidence in whether particular actors are 'doing a good job for society' with regard to biotechnology. Their most interesting finding is that Americans' trust in industry actors is on a par with consumer organisations, superior to trust in environmental organisations, and 26% above Europeans' average trust in industry actors (which remains at 32%). US respondents' confidence in farmers is also significantly higher (73% vs. 53% in Europe). Europeans give on average higher ratings to ethics committees, consumer organisations, and environmental NGOs. This pattern of responses is broadly consistent with Vogel's (2003) proposition that – because Americans' trust in national regulatory agencies is very strong and because society's faith in industry actors rose during the 1980s and 1990s – we should expect to see greater public acceptance and more permissive regulation of GMOs.[4] However, in recent US surveys (Lang and Hallman 2005; Lang 2013), no group of actors receives overwhelming confidence, while university scientists, farmers, and consumer and environmental groups are consistently rated more highly than government agencies, the media, food retailers, and food manufacturers. At the same time, even industry actors receive trust scores which are slightly more positive than a 'neither–nor' median value. Overall, the divergence between US and EU may by now be somewhat smaller than originally assumed.

Most studies on trust are informed by an underlying 'causal' model. They link trust in institutions or actors to overall societal acceptance of GM foods and crops. However, partial correlation does not amount to an overwhelming case for placing trust at the centre of attitude formation. It is therefore useful to also consider an alternative method for linking trust and acceptance. Drawing on UK survey data, Poortinga and Pidgeon (2005: 200) convincingly apply the 'associationist' model of trust. Whereas, in the 'causal' model, higher trust leads to more acceptance, in the 'associationist' model the relationship is reversed and 'the acceptability of a risk [... is] the determinant of trust.' These findings for the UK support the view that both trust and perceived risk are largely 'expressions of a more general attitude toward GM food' (ibid.: 206). In other words, there is a pre-existing structure of values and attitudes which helps amplify congruent information and create relations of trust with particular actors. The more these actors are perceived as sharing similar values with the public, the greater their trustworthiness becomes (Earle and Cvetkovich 1995). Priest et al. (2003: 765) equally emphasise that citizens' judgements must be understood as 'inseparable' from the wider socio-cultural context.

The contested nature of public opinion

Given that the above approaches to risk perception – scientific literacy, media-fuelled panics, and social trust – all have limited validity for agbiotech, it needs to be asked what the public's 'expanded vocabulary of risk' (Wilkins 2001) actually entails. The *cultural* politics of agbiotech will be explored in subsequent chapters, but the remainder of the present chapter examines the socio-cultural drivers of public outrage in general terms. The first two approaches to risk perception – scientific literacy and moral panics – are grounded in insights from psychometric models which emphasise ignorance and spontaneous emotional 'gut reactions' such as fear (Slovic 1987; Sjöberg 2000), fuelled by the unfamiliarity of risks and media scare stories. The psychometric approach is also readily linked to bottom-up processes of attitude formation. Affective reactions prompt spontaneous decisions about GM products and ultimately generate lasting opinions and beliefs about the technology.

However, there are problems with this model of risk perception. Sjöberg (2000) criticises its reliance on statistical averages and, in both US and European focus groups, a host of broader moral and political concerns was recorded (Marris et al. 2001; Wagner et al. 2001; Knight 2009). Especially in the European case, the term 'risk' is largely absent from focus group discussion, and the concepts of 'hazard' and 'danger' are frequently used in the context of moral debates about 'natural order' and related themes (Wagner et al. 2001: 85–86). Even in some statistical studies, risk does not always figure prominently. For instance, in their re-evaluation of the 2002 Eurobarometer, Pardo and Calvo (2006) found that risk perception did not influence the two leading motives for endorsing agbiotech – usefulness and moral acceptability. In the US, the discussion is similarly wide-ranging, but moral issues only come to dominate when the topic of animal-related biotechnology is broached (Knight 2009).

For European citizens, it seems best to theorise a top-down process of attitude formation whereby 'causality flows from values through attitudes to behavior' (Dreezens et al. 2005: 116). Although there is relatively widespread agreement on this dynamic in the literature on public attitudes, there remain some variations. Sjöberg (2000) holds that specific, identifiable values – namely opposition to 'interference with nature' – decisively shape attitudes towards agbiotech, and Ceccoli and Hixon (2012) claim that particular policy preferences – for organic agriculture and active consumer protection – are important. Most other scholars argue that overarching meta-attitudes (Costa-Font and Gil 2012) or general attitudes and values (Grunert et al. 2003) are the central causal factors. Whereas a bottom-up model of attitude formation expects citizens to consider the risks and benefits of each new biotech application on a case-by-case basis and then gradually arrive at an overall conclusion, a top-down process proceeds from 'higher-order life values' (Bredahl 2001) which then guide the evaluation of specific objects. In this vein, Pardo and Calvo (2006) interpreted the 2002 Eurobarometer as showing that people use a 'single evaluative mechanism' for all agbiotech applications, based on overarching worldviews. Due to the embeddedness of attitudes in broader belief systems, this also suggests that opinions are not easily changed by new information or novel applications of agbiotech (Bredahl 2001). For US citizens, higher order values also play an important role in attitude formation (Knight 2009), but given the dearth of qualitative studies, their nature and intensity remains underspecified. By contrast, there is considerable evidence for bottom-up attitude formation. Heiman and Zilberman (2011) demonstrate that negative framing lowers the likelihood of choosing a GM product, while a price discount caused the opposite effect. When measured in terms of monetary compensation, Lusk et al. (2006) found that English and French consumers demanded significant larger discounts to choose GM foods than US consumers.

But where does all this leave the dominant approaches to risk perception, and especially the regulatory focus on physical risks which has attained something of 'hegemonic status' (Seifert 2004)? The concept of risk may be familiar terrain for natural scientists and policy-makers, but it is less central for explaining the outpouring of popular concern (in Europe) because it is rarely the main motive for hostility towards GMOs. It appears as if the public is not particularly interested in the scientific debate – unless demonstrable hazards are indeed uncovered – but rather in the normative beliefs behind expert advice and political decisions – as well as the moral gap separating regulatory bodies and the public (Collins and Pinch 1998). Risk has nonetheless become a focal point in the political controversy due to the following reasons: first, regulators are drawn to its technical framing; second, (for GM-sceptics) the dominance of scientific discourse makes 'invoking an uncertain future [...] a comfortable way to voice non-specific concerns' (Wagner et al. 2001: 85). Risk has, in other words, become a 'catch-all category' and the public's unease 'finds a legitimate expression in risk' (van den Daele 2007). Numerous surveys have demonstrated that the strongest motives for public scepticism are moral concerns. Most consumer and environmental organisations have only sometimes campaigned on a moral or cultural platform – for fear of being ignored by policy-makers – and have tended to emphasise the potential health and environmental consequences of GMOs (Lassen and Jamison 2001). That is, they have typically conformed to the dominant risk discourse.

Ethics, morality, culture

The last section of this chapter is thus concerned with specifying and discussing the moral and cultural characteristics which are implied by top-down processes of attitude formation. In what follows, I assume a tentative distinction between the more rigorous and conscious domain of ethics and the realm of morality – informed by deeply, but less consciously, held values and identities. Both are strongly associated with the political debate over agbiotech, but, as will become clear, the broader sphere of morality is more central to public opinion. Furthermore, I concentrate on the role of ethics/morality in the American context because – as subsequent chapters will demonstrate – values, identities, worldviews, and moral concerns or even 'vetoes' lie at the heart of European anxieties.

Ethics

Ethics features prominently in public and academic debates, ranging from secular or religious scepticism regarding the instrumentalisation of nature/life to socio-economic and agro-ecological critiques of patented property rights, corporate dominance and agricultural intensification. To a modest degree, ethics has also been integrated into European policy-making. For instance, EU Directive 2001/18 on deliberate release of GMOs provides for consultation with relevant ethics committees. Furthermore, the European Parliament has long taken an interest in the ethical ramifications of biotechnology (Daviter 2009). And finally, the recent proposals for regulatory reform explicitly foresee the possibility of member states banning particular GM crops on 'non-scientific' grounds, such as ethical or cultural concerns.

In the US, ethical concerns have been systematically side-lined by a regulatory system focused on the concept of 'sound science'. They have, however, been taken up by counter-cultural movements and professional ethicists (Buttel 2005; Ramjoué 2007). Moreover, the regulatory system's veneer of natural-scientific neutrality is somewhat suspect, as 'regulatory science' is ultimately a product of compromise and cross-fertilisation between empirical and socio-cultural factors (Jasanoff 2005). Any regulatory decision is essentially also a form of *political* risk assessment and a value-based search for culturally acceptable degrees of risk (Scherzberg 2006). Furthermore, the 'politicisation' of science may also involve indirect mechanisms such as the allocation of funding for desirable research directions. Most often, however, socio-cultural values and political orientations 'seep' into the work of scientists and regulators by shaping their personal values and institutional priorities, prompting them to adapt criteria for evidence of safety, reconsider the place of scientific uncertainty, or even allow for extra-scientific input from civil society (Levidow and Murphy 2003). The precautionary principle which informs European environmental policy-making is an illustration of how cultural and political dynamics can shape scientific criteria and regulatory activities.

On both sides of the Atlantic, the predominant means of injecting ethical considerations into the political debate have been discourses of professional ethics that assess the new technology through logically rigorous analysis. Jefferson (2006), for instance, draws on the distinction between extrinsic and intrinsic objections to GM technology. Proponents of agbiotech generally make use of 'extrinsic' utilitarian ethics to estimate its usefulness – both in terms of increased economic productivity and the improvement of human lives. 'Intrinsic' objections are categorised as principled opposition to the 'unnatural' modification of plants, animals, and foods. Jefferson's conceptualisation is a classic example of how non-utilitarian qualms are tainted with the scent of 'illogicality', since the concept of the 'natural' is commonly known as the 'naturalistic fallacy' in professional ethics (also see Chapter 3). Jefferson's investigation thus produces a cautious consequentialism that highlights the aspects of necessity, safety, and effectiveness. These are all facets of technological innovation which fit with a utilitarian framework and conventional risk–benefit analysis.

An alternative approach is offered by Thompson (1997) who asserts that 'food biotechnology violates procedural protections of consumer sovereignty and religious liberty.' The first imperative – consumer sovereignty – is conceptualised as the 'right to know' and portrayed as an extension of standard democratic practices. Second, there may be religious reasons to avoid GM foods, such as dietary rules, the profanation of the sacred, or violation of 'Natural Law' (Hallman 2000). In essence, Thompson makes a subjectivist ethical case which is designed to protect people's liberty of conscience without relying on scientific and consequentialist evaluations. His first ethical imperative of consumer sovereignty may have struck a chord with Americans' values, as the recent proliferation of campaigns for GMO labelling in the US attests. Yet, regarding the potential impact of religious belief, the non-partisan Pew Initiative (2006) commissioned a series of opinion surveys and reasoned that '[w]hile religious belief has some impact, it is not a key source of variation in public attitudes toward biotechnology.'

Morality and usefulness

Most US studies play down the impact of moral or ethical qualms. Anderson et al. (2006: 189) report that there were no 'major ethical objections to GM food' in their sample of US consumers: 38% (vs. 19%) agreed that it is morally right to use GM technology to improve crop production. Similarly, Hoban and Kendall (1992) stated that merely 24% of consumers were morally opposed to modifying plants and 53% refused animal biotechnology. The much greater significance of the moral implications of animal biotechnology is corroborated by more recent focus groups from Arkansas (Knight 2009).

Apart from a somewhat weaker significance of moral qualms in the US, there often appears to be a disjuncture between ethical/moral concerns and actual consumer behaviour (Cormick 2005). This suggests the possibility that utilitarian–consequentialist thinking routinely crowds out other moral anxieties. What are essentially utilitarian concerns for human health and the environment may represent the dominant ideational force in American society and culture. More principled or categorical beliefs have a tendency to wither away when confronted with substantial and measurable benefits, leading to a gap between abstract moral injunctions and concrete judgements on particular products. Schilling et al. (2002) thus reported that a GM grass that requires less frequent mowing was deemed acceptable by more than three-quarters of Americans, although explicit support for GMOs only stood at 58%. Likewise, GM sheep producing medicines and vaccines were endorsed by 59% of respondents, despite only 28% consenting to the general concept of GM animals. In short, opinions become more crystallised and trigger a readjustment of attitudes towards GM applications when the potential benefits are presented (Hallman 2000; IFIC 2006).

Earlier Eurobarometer surveys equally suggested that, alongside a 'moral veto' on GM applications, there exists a 'usefulness' imperative which ensures that innovations with little perceived utility are rejected (Knowles 2001). Gaskell et al. (2004: 187) noted that many of the sceptical focus group participants highlighted the 'absence of perceived benefits and the possibility of non-GM alternatives to achieve similar ends.' Gaskell et al. found that the question of physical risks only began to play a role once a sufficient degree of usefulness had been acknowledged. This is not to be seen as a reaffirmation of a traditional risk/cost–benefit analysis, and it does not lend itself to trade-offs either. Rather, this pattern implies a highly specific, almost categorical focus on the underlying purpose and benefits derived from the new technology. Sceptics thus often doubt whether GM crops will genuinely help poor farmers in the global South, question its alleged benefits (such as higher yields, lower input costs), and emphasise commercial dependency on corporations.

Paradoxically, experimental studies which exposed consumers in four European countries to information outlining the benefits of GM products (environmental protection, lower price, and better health/nutrition) decreased participants' willingness to purchase such products (Frewer et al. 2003). Although this may seem counterintuitive, both of these research findings essentially confirm a top-down process of attitude formation which is largely – and sometimes entirely – unaffected by the framing of particular product innovations. This can only work, of course, if usefulness is not understood as part of a bottom-up, product-specific process of attitude formation, but as one facet of a deeper matrix of values and attitudes. Given the meanings of 'usefulness' that have emerged from European focus groups and their tenuous connection to product characteristics, usefulness should therefore be regarded as a non-economic concept with strong ethical overtones – such as broader societal usefulness, often linked to environmental integrity, public health, and international development objectives. Evidence from Denmark (Lassen et al. 2002) and Germany (Peters et al. 2007) further suggests that individualist, utilitarian considerations (e.g. better nutrition, flavour, lower price) tend to be much less important in many European countries than in the US.

Across the Atlantic, the usefulness of GM products was frequently related to 'healthier' alternatives for existing foods as well as to price reductions and environmental benefits. Here, 31.8% of US respondents (vs. 30.2%) subscribed to the statement that 'GM lowers cost, therefore it is okay' (Anderson et al. 2006: 191) and 'an unexpectedly large percentage of respondents' believed in significant environmental benefits from agbiotech. A brief comparison with attitudes in other countries is instructive. Hoban (2004) discussed the findings of a 2001 study in which consumers were presented with food products that had higher nutritional value. They were given the choice to stop buying the product once they were told that it was genetically modified. A majority of US consumers decided to continue using the product, while a majority of Europeans declined the offer. In another study, Lusk and Rozan (2005) contrasted American consumers' readiness to pay a 38% price premium for 'Golden Rice' (containing more vitamin A) with French consumers who required an almost 100% price reduction.

In line with such moderately pro-GM tendencies, several research conclusions indicate significant public concerns in the US about the potential costs of mandatory GM labelling. Attempting to estimate consumers' willingness to pay for additional labels, Hoban (2004) reports that only 17% of respondents were ready to pay an extra $10 on top of their annual shopping bill – a very small sum. Such cost-averse reactions were instrumental in several unsuccessful campaigns to introduce mandatory labelling at state or local levels. In Oregon, a 2002 ballot was rejected by almost three-quarters of voters. A major reason for this failure was the argument about rising food prices through onerous labelling requirements which was heavily advertised by a coalition of seed companies and food manufacturers (Toke 2004: 127). In November 2012, a similar ballot in California (Proposition 37) was narrowly defeated by a 51–49 margin, but this has led to similar legislative initiatives in over 40 US states. In summary, even when US citizens take an interest in issues of labelling, food safety, and ethical/moral issues related to agbiotech, their cognitive threshold for being persuaded by potential product-related benefits appears to be lower than that of Europeans. The latter will more frequently decide on the basis of higher order values and attitudes.

Conclusion

When surveying the existing literature, it is evident that the transatlantic divide over agbiotech regulation remains a productive research puzzle. Political scientists have applied a whole range of theories to the task and the insights generated by policy analysis and regulatory studies have often proved essential. By drawing attention to political opportunity structures, regulatory traditions and principles and by proposing a historically based theory of regulatory change and stability, institutionalist approaches make an important contribution. Yet, they have focused less on wider state–society relations and arguably underestimate the extent to which, particularly in Europe, regulatory reform has been driven by public opinion and political mobilisation. Political-economic analysis can be employed to consider a wider array of actors and highlight the importance of commodity chains as a parallel locus of contestation over agbiotech. These accounts often appreciate that it was, above all, citizen mobilisation and consumer power which led to the elimination of GM products from European commodity chains and pushed some member states, as well as the European Parliament, to initiate a legislative transformation. In the US, the relative weakness of agbiotech challengers and a relative lack of sustained 'public outrage' have allowed commodity chains and institutional structures to maintain their established routines for the time being.

A combination of institutionalist and political economy perspectives delivers a highly convincing, multi-causal explanation of transatlantic regulatory differences (e.g. Bernauer 2003; Pollack and Shaffer 2009). But to explain the intensity of Europeans' hostility to GMOs, scholars tend to either rely on risk–benefit analysis or on the impact of a wider regulatory crisis (and loss of public faith) triggered by BSE. While this is a crucial insight, it does not adequately explain why Europeans' concerns about agbiotech were tangible well before the 'years of controversy' and have not significantly abated until today. An interpretive perspective, by contrast, would argue that GM foods and crops have become a 'sounding board' for a host of moral, political, and cultural issues in Europe while rarely doing so in the US. Multi-causal explanations of the transatlantic divide sometimes acknowledge the relevance of context and of cultural factors, but they tend to regard them as secondary (Lynch and Vogel 2001), merely accord them the status of enabling conditions alongside contingent events (Pollack and Shaffer 2009), or treat them as one of several significant factors (Prakash and Kollman 2003; Montpetit and Rouillard 2008).

However, if public opinion and consumer power are critical drivers (and perhaps even the primary cause) of transatlantic regulatory divergence, then it is difficult to avoid the conclusion that the extraordinary mobilisation witnessed in Europe – and the relative lack of such pressure in the US – has, to a significant extent, been related to cultural values and identities. Naturally, cultural contexts do not directly *cause* specific outcomes, but they are highly influential in shaping the degree of public support for GMOs by offering particular cultural-political opportunities and foreclosing alternatives. Agent-centric explanations of agbiotech policy-making, which typically play down the relevance of cultural factors, struggle to clarify how and why political pressure has developed and why particular narratives and justifications resonate, while others fall flat. In reaching this conclusion about the deeper sources of public unease and 'moral outrage', this chapter has followed an analytical journey from conventional risk–benefit assessment to attitudes and values based on ethical/moral considerations. Because they fit well with the diagnosis of an 'irrational' European public and an event-based narrative of European opposition (triggered by the BSE crisis), the first two explanations – scientific ignorance as well as media-fuelled panics – have risen to prominence in both academia and the policy-making world. On balance, however, there is limited evidence that these two approaches offer a comprehensive explanation of European resistance to and relative US tolerance of agbiotech.

With regard to 'exogenous shocks' (such as regulatory crises), most perspectives do not fully account for why earlier scandals – such as the StarLink episode in the US or the discovery of dioxin-contaminated foods in Europe – have not aroused similar passions. By contrast, interpretive perspectives can be used to highlight the cultural resonance of particular storylines and the strength of pre-existing trends and tensions (Toke 2004: 42–43). Supporters of the scientific literacy thesis often assume fairly identical consumer preferences across countries and expect citizens to perform a risk–benefit analysis of new product innovations. If the outcome is negative, the causes are assumed to be a lack of accurate information, a predominance of misleading claims (fuelled by the media), or the absence of significant benefits for consumers. To understand European dynamics and, by corollary, relative US stability, it is better to think in terms of different rationalities (Wilkins 2001).

Social trust is another explanatory factor that can be part of a more interpretive understanding of risk. While the causal model of trust only produced modest correlations, the associationist model – which focuses on the congruence of values held by the public and by key political actors – can be linked to Europeans' preoccupation with the moral acceptability of GMOs. In the US case, the picture is less clear, with institutional trust (in regulatory agencies) at comparatively high levels and with moral concerns present alongside more individualist, utilitarian motivations. While the latter suggest a bottom-up dynamic of attitude formation, this finding on its own does not explain the origin of norms and values that support these utilitarian calculations. Moreover, how successfully are these values then 'activated' on both sides of the Atlantic by enterprising NGOs or status quo-oriented industry coalitions to push for or prevent regulatory reform? Finding answers to these questions is a central purpose of this book. An attempt to grasp the underlying moral dynamics of public opinion will therefore proceed through an analysis of the relevant cultural contexts in the US and Europe, including prevalent values and cultural identities.

The existing literature on the transatlantic *cultural* divide over agbiotech is mainly rooted in socio-legal studies, geography, and anthropology. Krenzler and MacGregor (2000), for instance, posit that laws and regulations evolve in response to changing conceptions of public policy which – even if translated through competing interest groups – can be traced back to different cultural attitudes. Similarly, Echols (1998) draws attention to European culinary traditions originating in the Middle Ages and theorises that '[c]ulture and tradition play a silent role in the regulatory process and the resulting rules.' Knowles (2001: 73) writes that 'reaction to GM foods in Europe reflects the history and cultural, societal and ethical values of European citizens', and Herrick (2005: 286) muses that the transatlantic rift 'can be theorized as the discursive representation of cultural, political, and economic identities'. The most notable exception to the relative neglect of cultural factors in regulatory studies are contributions by Jasanoff (2005), Kurzer and Cooper (2007), and Schurman and Munro (2010). They manage to weave together elements from comparative politics, political economy, and sociological analysis, thus creating powerful explanations not only of which actors were influential but also of why they became active and how they did so successfully.

Admittedly, generalisations at the scale of countries or even continents are always problematic, particularly in the light of internal American and European differences. Yet, they remain important heuristic tools to identify the potential driving forces or structural foundations of divergent transatlantic patterns. Building on the many existing insights reviewed above, subsequent chapters explore the cultural politics of agbiotech, with special attention not only to political dynamics, cultural values, and identities in the US and in Europe but also to their historical embeddedness and associated structural characteristics.

3
Theorising Culture and Nature

Introduction

Having identified public opinion and ethical or moral concerns about agbiotech as important elements for explaining transatlantic regulatory divergence, it is now time to develop the broader cultural-political approach. Cultural factors constitute the explanatory core of this book, while history serves as an analytical method to demonstrate the persistent relevance of cultural values and identities. This task is complicated by the fact that the concept of culture is one of the most contested terrains in the social sciences. Raymond Williams (1976: 87) famously judged that '[c]ulture is one of the two or three most complicated words in the English language', although, for many decades, disciplines such as history or sociology have put the concept to good use. In political science, however, culture has remained an under-theorised subject (Reeves 2004). This neglect of cultural analysis might well have ended with the controversial 'clash of civilisations' thesis by Samuel Huntington (1996). Yet, perhaps this flurry of interest does not constitute a genuine break with the past. As Kratochwil (1996: 203) notes, '[f]ar from representing a mere personal preference [...], questions of culture and identity have always been part and parcel of our analysis of the social world.' He diagnoses a degree of 'amnesia' in recent times which has prevented a more widespread use of cultural concepts.

Thus, to some extent, the cultural-political perspective of this book is symptomatic of a broader intellectual shift in the social sciences. Post-positivism has encouraged academics to take interpretive analysis, norms, values, and discourses more seriously. Culture represents one broad theme in whose environment alternative approaches can thrive. In this chapter, I not only draw on culture's associations of identity and belonging and its references to habits and traditions. I also propose an even deeper layer of cultural structures which can be labelled 'civilisation'. In the following, I begin by setting out a theory of cultural materialism in order to combine idealist and materialist strands of analysis. After elaborating the concepts of cultural context and cultural identity, I suggest that they (as well as discursive resonance) offer a way to partially reconcile the study of politics with cultural factors and civilisational 'background' conditions (which I call dispositions). In the second part of this chapter, I outline a cultural analysis of the agbiotech controversy and trace the importance of the concept of nature (and 'unnaturalness') for European reactions to the new technology. This prepares the ground for an analysis of contemporary regulatory politics in Chapter 4.

Culture and civilisation: A complex history

Some of the confusion around the notions of culture and civilisation arguably stems from their chequered historical evolution. The fact that culture is used both as a descriptive term for a way of life and as a label for the products of 'high' culture, art, and even education is not simply a matter of conceptual imprecision. Having its roots in Latin, culture is derived from the word *colere* which refers to activities of caring, tending to, and cultivating. The practice of 'agriculture' is nowadays closest in meaning, and, for the Romans, it was intimately connected to cultural practices (Williams 1976; Eagleton 2000). In the seventeenth and eighteenth centuries, the terms of culture and civilisation assumed connotations that are more easily recognised from a contemporary vantage point. Civilisation came to signify both 'a state of social order and refinement' and a more procedural sense of 'secular and progressive human self-development' (Williams 1976: 57–58). The notion of culture was less prominent at the time, but it tended to denote a similar certainty of social order, good manners, and superior knowledge.

At the end of the eighteenth century, however, the philosopher Johann Gottfried Herder introduced what is now regarded as the 'anthropological' concept of culture (Reeves 2004). Herder rejected the Enlightenment faith in a (peculiarly Euro-centric) evolutionary logic of human history. His intervention ushered in the pluralisation of the concept: culture became cultures. It is this sense of culture as a distinct way of life, defended by the Romantic movement and popularised by twentieth-century 'relativist' strands of anthropology, which springs to mind when engaging in comparative cultural analysis. Consciously defining itself against civilisation, the concept of culture began to turn into an explicit critique of modern industrial progress, especially in the German use of *Kultur* (Eagleton 2000: 10). However, the tendency of *Kultur* not only to celebrate cultural diversity but also individual refinement and self-development (as opposed to mere material advancement) means that – even today – culture as a way of life is not the only available usage (e.g. Elias 1978).

Of course, evoking the complex nature of culture and civilisation can only be a starting point. A first choice entails a focus on the anthropological meaning of culture. Regarding cultures (in the plural) as ways of life and modes of thinking reflects a certain appreciation of diversity and underscores the complexity of political and ethical judgement in the late modern age. Second, one should take account of the existing, ongoing philosophical debates about the meanings of the term 'culture'. A straightforward opposition between nature and culture – the nature–nurture debate – is not sufficiently exhaustive. In keeping with the Latin origin, Eagleton (2000: 2) suggests a median position between the natural and the constructed: 'If culture means the active tending of natural growth, then it suggests a dialectic between the artificial and the natural, what we do to the world and what the world does to us.'

Cultural materialism

As I understand it here, cultural materialism[1] subscribes to the tenet that human reality is socially constructed and is thus a variant of social constructivism. However, it does not support the idea that reality consists of 'ideas all the way down' and that we can entirely neglect our material environment. One should not give up on the idea of a material world outside of cultural constructions, even if human beings do not enjoy unmediated access to this 'material reality' (Jacobsen 2003). As Eagleton (2000: 4–5) puts it,

> The idea of culture, then, signifies a double refusal: of organic determinism on the one hand, and of the autonomy of spirit on the other. It is a rebuff to both naturalism and idealism, insisting against the former that there is that within nature which exceeds and undoes it, and against idealism that even the most high-minded human agency has its humble roots in our biology and natural environment.

Cultural materialism, as I have labelled it, thus occupies the middle ground between determinism and idealism. In this way, it can even act as a bridge that allows for a dialectical usage of these two extremes. One example is the ontological status of nature in this book – that is, the degree to which nature is physically and conceptually separated from the human realm. This partition is the result of both material/geographical and ideational factors and, in turn, has markedly influenced the cultural evolution of societies. As Chapter 5 will show, once American agriculture had been mechanised and sharply separated from both human habitation and 'wilderness', more integrative conceptions found it difficult to attract public support. I will theorise these structural features through the concepts of 'ontological condition' and 'civilisational disposition'.

A workable interpretation of cultural materialism equally occupies a median position on a social-constructivist continuum between voluntarism and essentialism. In line with social constructivism, ingrained cultural practices can be regarded as reproducing and thus securing social constructs against radical change. Essentialism, in this context, refers to the reification of social categories such as race, gender, or culture. An essentialist view of such categories allocates certain fundamental and unchanging qualities to them and sets up rigid distinctions with other categories. Cultures, on this view, are static entities and not dynamic processes. By contrast, voluntarism celebrates the human capacity of willing or choosing and subordinates more unconscious factors of human behaviour. It is thus directly opposed to what may be seen as reprehensible elements of habit and tradition inherent in cultural practices.

For the purposes of this book, cultural materialism represents a theory of the 'middle ground' and contains a broad spectrum of positions on the social-constructivist continuum. It is thus possible to hold that 'man is an animal suspended in webs of significance he himself has spun' (Geertz 1993: 5) without denying the essentialist argument that these webs can be extraordinarily powerful. In line with the voluntarist view, it is also true that human beings are able to reflectively challenge received wisdom and transform the dominant intersubjective understandings. Overall, nonetheless, human beings cannot escape the fate of culture. They cannot stop producing and reproducing culture, but they can make its workings more or less transparent and its effects more or less contestable.

The concept of civilisation

The term 'civilisation' has equally undergone a complex evolution. Representing ideas about 'cultivated' manners, Westernisation and Enlightenment, and an antithesis of 'primitivist' barbarism – meanings that continue to function in the present[2] – it began to be widely used by historians in the nineteenth century to portray vast socio-cultural structures of human organisation, such as empires or more loosely geographically connected, but culturally similar collections of peoples. In one sense, civilisation could be seen as a diluted and broadened concept of culture, but this framing downplays the distinct emphasis on long historical cycles of interaction between ideas and material foundations of human societies. Civilisations could therefore be thought of as an expression of cultural materialism. The civilisational theory of Robert Cox (2002), for instance, is situated within both cultural and historical materialism. He steers the concept of civilisation away from both the sphere of identity and from overly geographical accounts. For him, 'a working definition of civilization can be a fit or correspondence between material conditions of existence and inter-subjective meanings' (ibid.: 161). It is distinct from the concept of culture, as the latter tends to refer to smaller collections of people. In contrast to Huntington (1996) – who views civilisations as 'the highest cultural grouping of people and the broadest level of cultural identity' – Cox does not regard civilisations as a 'very large realm of identity'. Rather, the term conceptualises the intuitive, intersubjective understandings of a large group of people, operating at 'a deeper level of consciousness – a level at which something that has been shaped by the historical development of a people comes to be understood by them as universal and natural' (Cox 2002: 163).

Among those habitual understandings are general dispositions such as an emphasis on individual or community, on time or space, perspectives on spirituality and on nature. Articulating a strongly sociological theory of civilisation, Cox is sceptical of Fernand Braudel's (1995) somewhat spatial view of civilisations which is widespread in world history. Cox (2002: 165) argues that civilisations do not require well-defined geographical boundaries:

> Nowadays, it makes more sense to think of a civilization as a community of thought, [...] while acknowledging that inter-subjective meanings evolve in relation to material conditions in which geography continues to play a role alongside transnational economic networks and world-spanning communications technologies.

Although his view has much relevance in the late modern age with its world-spanning cultural flows and globalising economic elites, exploring the cultural politics of agbiotech cannot do without a stronger role of identity and physical geography. Regarding the former, I introduce a second level of analysis below. With respect to the latter, I side with Braudel's approach to civilisations. For him, individuals, societies, and civilisations are all part of a long historical process that is fashioned by both material and ideational factors. This process is called the *longue durée*[3], and civilisations are its prime illustration: 'Civilization is in fact the longest story of all' (Braudel 1995: 34).

Material factors are vital to explain a civilisation's ability to sustain itself, grow and contract, trade and wage war, evolve or disappear – but so are ideas and habits. The geographical conditions that allowed American frontier lifestyles to thrive, which I review in Chapter 5, were not sufficient on their own to trigger the same kind of settler 'invasion' in the uncharted territories of western Canada, where developments lacked comparable pace and intensity. Similar to Cox's (2002) emphasis on intersubjectivity, Braudel presumes that (what I label) 'civilisational dispositions' represent structures which are so deeply rooted that most people are unable to think or act outside of them: 'These are the "foundations", the underlying structures of civilizations: religious beliefs, for instance, or a timeless peasantry, or attitudes to death, work, pleasure and family life' (Braudel 1995: 28).

Of course, these structures can still be transformed in the longer term. But for short- and medium-range historical analysis, changes appear to be imperceptibly slow – at least until the rapid modernisation experienced in the twentieth century (Bess 2003). It is important to emphasise that the notion of civilisation – not least due to its abstractness – is a heuristic model which gives narrative structure and analytical focus to historical accounts. Naturally, there are plenty of peculiarities and contradictions to be found in every civilisational space around the world. This heightened indeterminacy of civilisation distances it from the notion of cultural identity, which I discuss in the next section.

This book does not seek to provide a comprehensive comparison of American and European civilisations. Given my focus on agbiotech, I confine myself to examining certain civilisational aspects (such as attitudes to nature, agriculture, and food). The first basis of this study is the divergent 'ontological condition' of European and American societies – the way that their material mode of existence displays either more interactive or more bifurcationist tendencies with regard to the natural environment. This ontological condition is then partly responsible for generating compatible civilisational dispositions which, in turn, serve to strengthen and reproduce the mode of existence through a continuous, recursive process. Exceptional historical transformations, such as modern industrialisation or the Enlightenment movement, can be seen as having thrown this relationship off balance because the material capabilities of most societies now greatly exceed their ideational ability to make sense of the world around them – and particularly of modern technological innovations like agbiotech.

Historical forces have typically advanced in an amalgamated fashion. What Friedman (1994: 50) calls 'early materialism', animated by the ideal of progress, is an illustration of this dual movement of ideas/values and material capabilities. A contemporary example would be the vastly expanded potential of the material-economic realm and ideas of 'bio-liberation' from ordinary human limits through rapid progress in biotechnology (Bailey 2005), but this new techno-cultural amalgam does not yet appeal to most people, especially not in Europe. The struggle between such new techno-scientific movements and deeply rooted attachments to long-standing cosmologies, ideals, and settled lifeworlds can be explored especially well through the notion of cultural identity.

Cultural identity and contextualism

While material structures and cultural context do not fully determine our actions, they effectively constrain the options for intentional change. Contextualism does not entirely exclude the possibility of transcultural moral discourses, moral intuitions, and concepts of justice (Scharfstein 1989; Benhabib 2002; Lawson 2006). But conversely, the observation that cultural boundaries are ultimately blurred and differences historically constructed does not mean that cultural aspects can be neglected. The fact that dominant meanings and values are often contested by 'counter-cultural' individuals or organised groups does not obviate the existence of a shared social context with mutually intelligible meanings and practices (Ross 1997). In this vein, despite advocating transformative, deliberative political practices, Benhabib (2002: 7) argues that social constructivism 'does not suggest that cultural differences are shallow or somehow unreal or "fictional". Cultural differences run very deep and are very real'.

Conceptualising cultural differences

Taking its cues from early twentieth-century anthropology, essentialism treats cultures as rigid and sharply demarcated structures that almost entirely determine their members' lives. Whereas this view has some merits in an anthropological study of a remote and isolated society, it becomes untenable once the focus widens to civilisations. In contrast to Huntington's (1996) essentialist take on civilisations, Cox's (2002) approach emphasises that the notion of identity cannot be so easily deployed. Despite some scattered examples of fierce dedication to, for instance, an Islamic civilisational identity, the concept of civilisation is much more relevant to very basic structures, ways of life, and thought. The emotional attachment inherent in the notion of identity is best applied to more cohesive forms of culture.

Essentialist perspectives have difficulty in accounting for political and cultural change. While a certain measure of voluntarism is indeed necessary, it is tempting to overemphasise the political malleability of cultural structures. Reeves (2004), for instance, defends a modernist voluntarism that seeks to resurrect a universalist discourse. Reeves advances the case for a substantive universalism that takes its inspiration from the Arnoldian view of culture – 'the best that has been thought and said'. Much of her work is concerned with attacking the 'anthropological' particularist view of culture as a way of life by politicising the concept and exposing it as a contestable category: 'Culture is not a truth claim – it is a political statement and as such deserves to be handled with the same measure of intrigue, criticism, suspicion or contempt as any other claim' (ibid.: 183). The Arnoldian, humanist notion of culture could give rise to a universalist cosmopolitan discourse, providing a 'language of commonality, not difference' (ibid.: 184). In other words, Reeves proposes a dissolution of culture in the practices of politics. While there is much to be said for a perspective that sees culture and politics as inextricably entangled systems, the virtual abolition of 'thick' culture (signifying deep-rooted difference) is not in tune with an obstinately diverse world.

Cultural identity and the end of modernist hegemony

Given the drawbacks of both essentialist and voluntarist interpretations of cultural difference, an alternative perspective is needed. In his exploration of worldwide culturalist revivals, Friedman (1994) provides a suitable definition of 'cultural identity'. In its strongest essentialist or 'ethnic' variant, cultural identity denotes 'the attribution of a set of qualities to a given population' and is 'not practiced but inherent [in the "blood"], not achieved but ascribed' (ibid.: 29). At the opposite end of the spectrum lies the weakest, 'modernist' variant of cultural identity – closer to lifestyles or ways of life and with no direct basis in tradition. In between lies what Friedman sees as the 'general Western notion of ethnicity', a moderate version of attribution expressed as 'heritage' or 'cultural descent' that is passed on across generations. The latter two definitions are relevant to the politics of agbiotech because a culturally articulated opposition to GMOs can be found in both modern identity construction (e.g. through consumerist practices) and, more importantly, the neo-traditionalist emphasis on cultural heritage and authenticity.

The debate on the nature of 'culture as difference' and its tendency to slide towards essentialism is less central in this respect. Culture may sometimes be 'a relatively instable product of the practice of meaning' (Friedman 1994: 74). Yet, it does not follow that the theorist's task is confined to a critically minded labour of deconstruction, exposing relations of power and mechanisms of stabilisation. In this book, therefore, the understanding of historically fashioned differences and their effects on contemporary politics takes precedence over the exposition of political alternatives. As Friedman (1994: 238) notes,

> History, language and race are all possible bases for cultural identity and they are socially constructed realities. This does not make them false and ideological if we recognize the degree to which all identity is constructed. Identity is only false for those who have none or feel alienated enough from any particular identity that they could never dream of participating in such quasi-religious mystification.

Cultural identity constitutes a means by which people 'derive a sense of who they are, how they should act, and where they are going'. Identity is, in other words, the most prominent 'action unit of culture' (Fitzgerald and Campbell 2001).

Besides its analytical utility, cultural identity needs to be recognised more urgently than ever due to its contemporary resurgence. Friedman (1994: 78) observes 'an implosive loss of faith in the progress of "civilization", and a corresponding explosion of new cultural movements, from cults and religious revival to primitivism, a new traditionalism, a striving for the re-establishment of a new culturally defined identity'. As I will show, resistance to the agbiotech revolution as a symbol of the modernist utopia is a part of this dynamic. And even if cultural identity is to be the 'action unit', it also represents a political category when articulated and mobilised for concrete ends, as is often the case in the politics of agbiotech.

Culture and civilisation as context: Linking culture and politics

Contextualism evokes a concept of culture that spans the divide between determinism and idealism. Scharfstein (1989: 1) defines it as 'the study of the way in which contexts explain, or [...] the view that explanation is impossible or seriously incomplete unless context is taken into account'. More critically inclined, Lawson (2006: 42) argues that cultural contextualism, by emphasising or even reifying cultural traditions, stipulates an unbroken continuity between past, present, and future. This perspective would not only create a new foundationalism (culture-as-context) but also potentially 'freeze' dynamic cultural processes into a problematic structure. However, it is entirely possible to avoid following the structuralist path to its most radical conclusion. In this book, therefore, context denotes the central 'possibilist' idea that cultural limits to imagination and action are powerful, but not strictly determining the future. When dealing with constitutive and unconscious civilisational dispositions, the notion of contextualism hence corresponds closely to Cox's (2002) structure of intersubjectivity.

While I focus on cultural contextualism to illuminate the public reception and the regulation of GMOs in Europe and the US, the boundaries drawn with adjacent territories are never absolute or impermeable. Europe, as I argue below, is an analytical construct legitimated by a long history of similar humanity–nature relationships. As the concept of civilisation allows ample scope for internal contradictions, a high degree of cultural unity is not required to justify a general underlying scepticism towards the modernisation of intensely cultural spheres of life. The use of concepts such as cultural identity and cultural values complements the civilisational storyline by offering better-defined pathways through which the cultural politics of specific societies is played out. The idea of nature, for example, was fundamentally transformed over the course of American history and has nonetheless retained a certain property of separateness or distinctiveness with regard to the human sphere. A civilisational pattern or 'disposition' has thus been established, but nature can still be perceived in a number of different ways, for instance as either 'howling' and threatening or as 'sublime' and revered wilderness (see Chapter 5).

The question remains, nonetheless, of how to reconcile the voluntarist ideal of political discourse and action – the belief that we can shape our destiny – with slow historicist processes of civilisational change and the relative resilience of cultural practices. Some versions of discourse theory seek to clarify the linkages between political discourse and cultural context. O'Mahony and Skillington's (1999: 101) central claim is that 'innovations in the social construction of reality are never so transformative as to be culturally unrecognisable.' Against discursive voluntarism, they further insist that 'discourse formations do not so much complement wider systems of cultural or political belief as respond to them in a creative manner.' Put differently, political actors do not create new ideas or movements out of thin air, but creatively employ pre-existing attitudes and belief-systems for their ends. Many forms of discourse theory risk overstating the extent to which particular groups or the wider public can be influenced by innovative new frames. Recognising the significance of cultural contexts means that a strongly constructivist interpretation of culture is less suitable for uncovering deep-seated cultural values and motivations with regard to agbiotech. This caveat harks back to the introductory chapter where I used Mishler and Pollack's (2003) distinction between 'thick' (fundamental, exogenous, holistic) and 'thin' (constructivist, endogenous, individualist) interpretations of culture. There have been some attempts to synthesise different approaches to culture. For example, concepts such as cultural 'tool kit' (Swidler 1986) or 'repertoire' (Tilly 1978) seek to capture both the constraining power of cultural forces and their inherent malleability. Yet, they seem rather instrumentalist and strategic (Williams 2004) because they imply that political actors will know how to recognise and exploit cultural opportunities.

After all, the objectives of political actors are not freely conceived or mirror those of rational utility-maximisers. Even political elites cannot be regarded as existing entirely apart from their cultural context(s), although, in the longer term, they can help transform this background through reflective action. As Jackson (1999: 150) writes, it is vital to complement an instrumental approach, in which identities or ideas serve as tools, with a constitutive approach that recognises the cultural embeddedness of political agents. In this respect, culture and civilisation are at the heart of a theory of context that illuminates the circumstances within which ideas are conceived and strategies are elaborated. Politics, from simple bargaining processes to complex mobilisation of public opinion, is better conceived of as 'cultural politics', simultaneously drawing on meanings and values as well as material resources and interests. To a degree, any theory of context will always remain an unfinished business, since its breadth precludes well-defined rules of analysis or comparative testing procedures. Nor is cultural analysis synonymous with critical discourse analysis because the former focuses on how the broader context influences the effectiveness of discursive interventions. Due to their interest in political processes of stabilisation and change, critically inclined interactionist or Foucauldian versions of discourse analysis (Hajer 1995; Feindt and Oels 2005) tend to highlight the constitutive power of discourses and thereby de-emphasise the constraining impact of context. In line with O'Mahony and Skillington (1999), however, a cultural-political analysis would insist that resonant discourses or frames both *reflect* historically evolved values/interests and *reconstitute* them to create a slightly altered reality. Novel discourses that do not resonate will struggle to make an impact, unless they are accompanied by long-term efforts to build supportive cultural constituencies and political alliances.

Cultural resonance and power

Cultural resonance is one of the mechanisms in this book for operationalising the notion of cultural context. If culture – both as a system of meaning and as a background of dominant beliefs and values – structures a field of political possibilities, it is both a strategic 'tool kit' and a constitutive force. The most successful frames used by actors in the agbiotech debate are those that manage to mobilise large sections of society and turn latent opposition (or support) into solidified opinions or even active campaigning. A good example comes from the broader environmental literature. Buijs et al. (2011) demonstrate the effects of cultural resonance in their study of a local conflict over national park management in the Netherlands. An ecocentric discourse focused on nature as wilderness did not persuade most local citizens, while an anthropocentric discourse centred on aesthetic and functional aspects resonated much more strongly. The authors highlight the significance of the 'cultural background against which stakeholders frame a conflict' and argue that 'the possibility of any reframing effort is directly related to the strength of the cultural values and beliefs on which the frames are based' (ibid.: 339). They also note that political conflicts rooted in divergent cultural views about nature will likely be difficult to resolve.

Chapter 4 will look in more detail at how the anti-GMO movement in Europe managed to craft messages that resonated with the cultural values, beliefs, and identities of large numbers of people. Conversely, a similar degree of resonance was initially not achieved in the US, although the dominant discourses of physical risk and 'consumer sovereignty' are now beginning to translate into substantial public awareness and political activity. The study of cultural resonance should further be complemented with the concepts of credibility and salience (Benford and Snow 2000). Credibility is not only related to the internal consistency and empirical plausibility of a discursive frame, but also refers to the credibility of those articulating it. This aspect thus refers back to the notion of social trust which I explored in Chapter 2. Salience is linked to 'centrality' – how essential the values and beliefs articulated in frames are to the public – and also varies according to how much the chosen frames resonate with everyday life and personal experiences. In a European context, as explored below, the ideal of naturalness appears to have a relatively central role in people's 'hierarchy' of cultural values, while its manifestation in food and agriculture makes it an integral part of the 'lived' experience. By contrast, Williams (2004: 106) provides an excellent illustration of how a widespread cultural admiration for pristine wilderness in the US achieves a high degree of resonance but may lack sufficient salience when it comes to conflicting policy objectives. The preservation of wild nature is sometimes overcome by 'approaches to nature that subordinate it to human needs and view progress as the developing control of the natural world.'

When cultural resonance is used to explain the success and failure of particular discourses and associated movements, it may even function as a form of power. For instance, O'Mahony and Skillington (1999: 112) distinguish between the cultural and political power of discourse coalitions which have influenced the biotechnology policies of European countries. 'Such coalitions', they insist, 'could not exist without at least some measure of support in public culture'. This distinction is important because when the authors write that the anti-biotech discourse coalition has 'more cultural power than political influence', they also imply that mere resonance with a society's dominant values does not always determine the policies that are actually put in place. There are at least two possible ways in which political and economic elites can respond to a cultural context that is dissonant with their project. Apart from ignoring it – which remains an option in states with a weak or disinterested civil society – they can act as discursive entrepreneurs and attempt to re-inscribe their political objectives into a more palatable cultural narrative. For example, the agbiotech industry has (with some success) appropriated the charitable project of alleviating world hunger by touting GM seeds as the best solution for a growing world population, thus connecting biotechnology with the values of charity and solidarity as well as narratives of progress and technological salvation. Yet, O'Mahony and Skillington (1999: 113) also emphasise that the cultural power of the industry coalition is fragile and limited and that it is 'likely to be undermined anew as further scientific frontiers are crossed with immense practical and moral implications'. And indeed, surveys and focus group discussions on both sides of the Atlantic suggest that some new biotech applications, such as animal biotechnology, are bound to elicit even more intense moral concerns. For now, in Europe it seems that greater cultural power still resides in the concept of 'unnaturalness' and that it can be used to stigmatise agbiotech products. The second part of this chapter concentrates on the enduring, albeit complex, relevance of unnaturalness for understanding public attitudes in Europe.

The cultural politics of nature and its relevance for agbiotech

Conceptions of nature

The concept of nature has been subject to even more ambivalent usage than the notion of culture. Raymond Williams (1976: 219) famously called nature 'perhaps the most complex word in the language'. He extracted three important meanings of nature which have achieved widespread resonance throughout much of Western history. Nature denotes primarily '(i) the essential quality and character of something; (ii) the inherent force which directs either the world or human beings or both; (iii) the material world itself, taken as including or not including human beings'. Any number of combinations of these three connotations is entirely possible. The Romantic view of nature, for example, assembles elements of all three definitions: an intuition about essences and unadulterated authenticity, a sense of spiritual power and primal importance, and material nature/wilderness with (or without) human beings.

At the civilisational level of analysis, and initially closest to William's third category, I concentrate on the ontological status of nature vis-à-vis human populations and ask whether their ideational and material relationship is marked by stark separation or by relative interactivity. Chapter 5 illustrates that the demands of environmental sustainability in a finite material space have created strong pressures towards the latter tendency. Rather than making a claim about civilisational difference in all spheres of life, I target this judgement at a particular area. But I do assume that the 'ontological condition' of humanity–nature relationships strongly influences environmental politics in a given society. Finally, I also go beyond a materialist analysis by considering the continuous interaction between the material sphere and cultural 'images' of nature – an element that is also captured by Cox's reference to underlying structural patterns of thought and behaviour which 'operate at a deeper level of consciousness' (Cox 2002: 163). Civilisational dispositions and ontological (environmental) conditions lay the groundwork for diverse cultures and identities while still restricting the breadth of possible outcomes. In Europe, the dominant American 'environmental identity' of wilderness preservation is thus disadvantaged from the outset, although a few countries (such as Norway) have shown some signs of moving in this direction, in line with their own peculiar environmental histories (Witoszek 1997).

The multiple meanings of nature come into their own when engaging with the cultural politics of agbiotech. Once Williams' (1976) second definition of nature as a guiding 'cosmological' force becomes relevant, it is necessary to further distinguish its connotations, albeit without becoming too categorical about conceptual boundaries. To make 'nature' theoretically useful, I thus rely on the distinctions made by John Stuart Mill in the nineteenth century. As Sagoff (2004: 5) articulates them, nature represents:

(1) everything in the universe (natural vs. supernatural);
(2) creation in the sense of what God has made (sacred vs. profane);
(3) that which is independent of human influence or contrivance (pristine vs. artificial); and
(4) that which is authentic or true to itself (trustworthy and honest vs. illusory, superficial, and risky).

Finally, a fifth meaning of nature in late modern societies should be added. Cultural critics such as Raymond Williams (1973) or Lowenthal (1991) have consistently highlighted the use of nature as a vehicle for popular nostalgia about the old times, when lives were more wholesome, less technologically driven, landscapes visually more stable, and identities more settled. 'Nature as nostalgia' captures this sense of cultural opposition against the rapid transformation of people's lifeworlds, especially in an age of globalisation (Macnaghten and Urry 1998: 223).

I will attempt not to label public attitudes to nature (and GMOs) too schematically, as opinions, values, and emotions tend to work in complex and overlapping ways. Nevertheless, both Williams' and Mill's concepts of nature are very useful to illuminate certain positions and responses. The central divide between scientific/ethical/political experts on the one hand and the public and NGOs on the other hand concerns the refusal of the former to admit more than Mill's first sense of nature, which can be described as 'environmental instrumentalism' (Macnaghten and Urry 1998). This instrumentalist image of nature is also reflected in the 'product-based' US regulations that regard GM foods and crops as substantially equivalent because their bio-chemical properties are assumed to be completely analogous to non-GM foods and crops.

Origins and enduring significance

All of the alternative notions of nature are largely concealed from open debate because of the predominance of rationalist traditions of thought which, in turn, have been shaped by orthodox scientific and instrumentalist perspectives. To underline the divide between instrumental and cultural rationalities, subsequent chapters will focus on Williams' second and third notions of nature, demonstrating nature's cosmological appeal and material presence in human societies throughout the ages. Particularly the second notion can be extracted from the rich source material presented by European surveys and qualitative studies on GMOs. As Wagner et al. (2001: 92) note, '[t]he high degree of moral rejection pertaining to certain biotechnological applications [...] is primarily owing to the image of "nature as a spiritual force" .' This second meaning of nature has exerted a powerful influence on societies across the centuries.

Back in the Middle Ages, the study of nature was associated with finding the sources of (divinely ordained) morality and animals or plants were perceived as emblems of underlying essences (Coates 1998: 60). In the realm of 'common sense', popular theology, and philosophy, however, nature in all its manifestations – weather patterns, seasons, and 'natural' catastrophes – had powerful spiritual and practical tasks to fulfil. Not entirely coterminous with God, Nature had its own place in the wider cosmology of the time: God was regarded as primary and Nature as his 'minister or deputy' (Williams 1976: 221). In line with customary political regimes, Nature was also pictured as an 'absolute monarch' who would decide at will, often arbitrarily causing great suffering for humans or, with luck, bringing fortune and grace. Nature was neither exclusively loathed nor revered, but feared and praised in equal measure. Adapting to the 'natural order' laid down by God and administered by Nature was the sine qua non of medieval life (Macnaghten and Urry 1998: 10). A similar sense of 'natural order' has survived in popular consciousness and finds its most explicit contemporary expression in Roman Catholic theology.

The European Renaissance (14th–17th centuries) ushered in a gradual shift in perceptions. Nature was demoted to a 'constitutional monarch' subject to 'natural laws', mirroring political developments in many European countries. Another implication of this shift was the removal of nature's animistic credentials. The emphasis on discoverable laws turned a 'life-giving', if somewhat unpredictable, force into 'dead matter' and began the transition from 'spirit to machine' that defined the epoch of industrialisation (ibid.). Today, a spiritual understanding of nature is thought to be confined to the margins of society, residing only in the minds of strong religious believers and deep ecologists. But this is a misconception. Cultural and environmental historians have long pointed to the power of unconscious belief systems or 'folk memory' which manifest themselves as veritable cultural trajectories. As Worster (1993: 200) wrote about Protestantism in the US:

> Protestantism, like any religion, lays its hold on people's imagination in diverse, contradictory ways and that hold can be tenacious long after the explicit theology or doctrine has gone dead. Surely it cannot be surprising that in a culture deeply rooted in Protestantism, we should find ourselves speaking its language, expressing its temperament, even when we thought we were free of all that.

The amnesia concerning the second sense of nature has adverse effects on our understanding of contemporary environmental politics. This meaning is not sufficiently appreciated in the literature, as the past 200 years have seen the triumph of modern scientific rationality over ancient 'superstitions'. Not only were spiritual perspectives on nature delegitimised, but the material impact of nature/environment on human flourishing was also de-emphasised. In the Enlightenment mind, Europe – 'by virtue of its intellect and industry' – had taken a great leap by 'breaking free' from environmental determinism and colonising those peoples around the world still living under the yoke of nature (Arnold 1996: 25).

The Romantic movement partly arose in opposition to this downgrading of the natural, imagining it instead as an 'expressive realm of purity and moral power, to be enjoyed or worshipped' (Macnaghten and Urry 1998: 22) and inspiring the modern environmental movement. The objections to GMOs that are advanced by European publics comprise a complex amalgam of spiritual, socio-cultural, and 'post-normal' scientific considerations. The former two aspects form the core of the analysis below as well as in subsequent chapters, reflecting the perceived loss of transcendental meaning and the destabilisation of the lifeworld. The de-spiritualising consequences of modernity – what Schiller called the 'disgodding of nature' – are commonly captured by Max Weber's notion of the 'disenchantment of the world' to which I will return in later chapters. The lifeworld, emphasised in the philosophy of Habermas (1984), stands for the social background or context; one might call it the true 'habitat' of human beings. Often labelled 'everyday life' in much of Anglo-Saxon philosophy, and used interchangeably in this book, it is 'the communicative space of society' (Crossley 2002: 154), the place for quintessentially human processes: play and leisure, 'identity development, the transmission of cultural knowledge, the resolution of disagreements about ethical and normative matters' (Sloan 1999). The lifeworld is opposed to the 'system' of formal, instrumental rationality and efficiency objectives which have been reinforced by the rise of capitalist modernity to a point where the system begins to crowd out the vital functions of the lifeworld and, as Habermas puts it, 'colonises' it.

The common theme of such counter-modern reaction (both religious and secular) is the reassertion of cultural meaning against a rationalist cosmology. The civilisational point here is that both spiritual and sociocultural anxieties are grounded in older, and partially enduring, patterns of an interactive relationship with the broader environment, which I have described through the notions of ontological condition and civilisational disposition. In practice, when exploring public concerns in more detail, Williams' second and third senses of nature are often intertwined to the point where it becomes impossible to distinguish them. They manifest themselves in the dominant 'cultures of nature' of particular societies. In the remainder of this chapter, I limit myself to setting out a number of different approaches to nature and 'unnaturalness' and relate them both to (European) public opposition against agbiotech and to the above typologies.

The moral 'trump card' of unnaturalness

When confronting the charge of GMOs' potential 'unnaturalness', supporters of agbiotech often fall back on a straightforward opposition between scientific knowledge and irrational public fears or political campaigning. Appealing to 'reason' and 'sound science', some commentators insist that 'junk science' and 'prejudice' should not be allowed to prevail (Caulder 1998). Even those who recognise that cultural values may give rise to the 'emotive symbolism' of 'killer tomatoes' and 'winged vegetables' cannot conceal their disdain for the 'primal fear' and 'spiritual resentment' animating supposedly 'modern' societies (Scott 2000: 304). Other scholars, however, take issue with the belief that emotions are 'erred judgements about the world' (Solomon 2003: 31). Considering the 2001 outbreak of foot-and-mouth disease in the UK and the proposed vaccination of livestock (and concomitant public unease), Beekman (2006: 307) suggests that we should contemplate 'what these negative emotional responses tell us about people's moral values with respect to livestock production'. These insights could equally be applied to GM foods. Ultimately, emotions are not merely atavistic reactions such as fear of spiders or snakes. They are often 'judgements in which people acknowledge their neediness and incompleteness, faced with elements that they do not fully control' (ibid.: 306). Ethicists sometimes use the notion of the 'yuck factor' to describe such spontaneous judgements of disgust at the underlying process itself (Agar 2003). In the case of GM food, this might, for instance, relate to the breaking of ancient taboos – similar to the (far stronger) injunctions against cannibalism – about the mixing of 'naturally' separate species to produce chimeras (Cox 2007).

One of the most sophisticated positions within the pro-agbiotech camp was articulated in a 1999 report by the European Federation of Biotechnology (EFB). Like many other sceptical bio-ethicists (e.g. Cooley and Goreham 2004), the EFB questions the usefulness of nature as a yardstick for assessing the benefits and costs of emerging biotechnologies. Having identified a belief in the unnatural 'essence' of GM foods as the root cause of the public's concerns, the EFB (1999: 11) argues that people commit a 'naturalistic fallacy' by deducing 'what is morally right and wrong from certain facts about the world and about Nature.' It proposes instead to view nature as morally neutral, 'simply as a factual situation which confronts us', and concentrate on the consequences of biotech applications. Moreover, the EFB also engages deontological and religious interpretations of nature – for example, as the revelation of God's will – and points to 'scriptural support' for a different perspective in which humanity's dominion over nature elevates it to the status of co-creators with God. Overall, the EFB thus makes thoughtful contribution to the debate, especially if compared with the biotech industry's diffusion of one particular image of nature through its advertising strategy – that of humans giving nature 'a nudge towards greater efficiency' (Levidow and Carr 1997: 33).

Yet, the EFB still has an incomplete understanding of those moral intuitions that originate in pre-modern times – where the dominant authority of natural science had not yet taken hold. The charge of unnaturalness has a long history and often comprises a whole variety of popular concerns, not all of them based on religious faith or deontological reasoning. The strongest focal point of the second sense of nature, as alluded to in the historical account above, tends to be the commitment to a natural order. The notion itself can be traced all the way back to classical thought, but gained a central position in Christian theology, usually coupled with the 'great chain of being' which arranged natural and social worlds into a God-given hierarchy of stations. It constituted the 'common sense' of the Middle Ages, lingering on even today. One of the most explicit renderings of this stance came in the late 1990s from the Prince of Wales who warned that '[g]enetically altered food crops take mankind into realms that belong to God, and to God alone. It is wrong that nature has come to be regarded as a system that can be engineered for our own convenience' (quoted in Smith 2003). This statement starts with an explicitly religious foundation, but then expands to a notion of hubris which can also be secularised. Lassen et al. (2002: 264–265.) observed that European focus groups often appeal to the 'Order of Nature principle' without, however, regularly mentioning God. This points towards a contemporary usage of natural order which combines both religious and quasi-religious or secular accounts (Sjöberg 2008). The latter orientation may be especially strong among environmentalists, but the widespread use of the natural in advertising and the ubiquity of unnaturalness in public discourse suggest a much broader resonance.

Fundamentally, the charge of unnaturalness reflects the wish to preserve a sacred realm that should not be tampered with by human beings, 'a moral intuition that the realm of nature and the realm of art are in some sense distinct' (Kirkham 2006: 182). A similar analysis by Chapman (2005) typifies an environmentalist reaction to the question of unnaturalness. Because people commonly recognise differences between more or less natural products (e.g. cotton vs. polyester), Chapham arrives at a 'biocentric' view that celebrates the authenticity and autonomy of nature. This position is not explicitly religious but still transcends the human-centred approach of instrumentalism, thus appearing somewhat similar to Prince Charles' position: 'Natural processes and living organisms have their own autonomy, an "otherness" that is always, to some extent opaque to human understanding: nature "lives and grows by itself"' (ibid.: 86). These interrelated conceptions of nature thus cover the whole range of meanings proposed by J.S. Mill, except for the first one – everything is natural (and, by corollary, humans are co-creators of God) – which is defended by the agbiotech industry and sympathetic bio-ethicists. Taken together, sceptical attitudes give rise to a critique of human hubris which is believed to be responsible for interfering with sacred realms as well as tampering with life-giving processes that humans do not fully understand (Kirkham 2006). This latter aspect will be briefly discussed at the end of this chapter.

A second use of the unnaturalness argument relates to the 'wisdom tradition' of virtue ethics. This second layer of values and norms serves to regulate the more deontological quality of natural order principles.

Kirkham (ibid.) thus contends that the intentions of political agents themselves are the central objects of public scrutiny: 'When objections about unnaturalness or "playing God" are used they are best understood as relating to the nature of the agent rather than a "nature" conceived as something other than or apart from the agent.' Concerns about character or identity of the agent are captured by the causal model of social trust which was discussed in Chapter 2. Virtue ethics may, however, also allow for a more bottom-up process of attitude formation whereby GM applications with a positive moral purpose – such as vitamin-fortified 'Golden Rice', drought-resistant plants, or GM mosquitoes (unable to spread malaria) – do achieve higher rates of acceptance (Lassen et al. 2002). Important objectives might therefore allow for greater 'interference' in the natural order, for instance if motivated by virtues such as compassion and charity (Deane-Drummond et al. 2003). Nevertheless, the range of applications where the profit motive does not predominate may be rather limited, given the central role of private corporations in biotech innovation.

Virtue ethics highlights the fact that all spheres of life are fundamentally experienced and interpreted in moral terms. As Chapter 6 emphasises, current debates about GMOs build on this long-established pattern. Examining the significance of earlier peasant protests in England against enclosures, Schaffer (1997: 124) brilliantly expounds the similarity of past moral outrage and contemporary discontent:

> In the early modern as in the postmodern world, challenges to cultural order were often seen as threats to nature itself. Consider, for example, the revival in capitalist societies of historicist values which deny an Enlightenment faith in progress and the growth of social movements aimed at preserving a carefully defined national (and natural) heritage.

The boundaries between cultural and natural order are necessarily blurred. Following my conception of culture/civilisation as the *via media* between material determinism and humanistic idealism, 'cultures of nature' both materially organise nature–society relationships and allow human beings to make sense – cognitively and emotionally – of this interdependent existence. 'Cultures of nature' can thus be located alongside other forms of visceral identity politics, dedicated to the defence of traditional lifeworlds, long-established social practices, cultural values, and the cultural landscapes of production, leisure and aesthetic appreciation. Cultural critics such as Raymond Williams (1973) have often been uncomfortable with this perceived conservative reaction to socio-cultural change. Yet, Williams underestimated the scale of the changes that were afoot in late modernity. His critical analysis of 'nostalgia for a visual and social world which existed 30 years previously' could not have predicted the disruptive power of cultural globalisation, of technological quantum leaps, and of 'hyper-capitalism'. As the economic system was increasingly infiltrating people's lifeworld, among the main tools for defending the latter were ideals of nature and of humanity (Zwick 1998).

Productivism and industrialism, in both business and agriculture, were joined by the supreme authority of modern natural science, signalling an end to cultural constructions of the world that did not meet the standards of rigorous measurement and material usefulness. As Mary Douglas (1975) once mused, a sole legitimate perspective furnished by science is bound to extinguish all possibilities of meaning. This retrenchment of the world to a rationalist 'optimum', rather than specific worries of imminent physical danger, is what most fundamentally underlies public anxiety. Sagoff (2004: 10) aptly noted that '[g]enetic engineering poses a problem principally because it crosses moral, aesthetic, or cultural – not biological – boundaries.'

Societal responses to the implications of the gene revolution at times even reach beyond the defence of established identities towards a quest for recovering basic notions of stability and authenticity. The manipulation of the building blocks of life is perhaps only comparable to another watershed in the seventeenth century when natural science began to demonstrate that, cosmologically speaking, the earth (and its inhabitants) ceased to be at the centre of the universe. In the case of GMOs, the 'Copernican abolition of obviousness in modern life has provoked counter-reactions that seek to restore a sense of normality to the world.' In response, '[t]he normal (i.e. what is taken for granted) is equated with the normative' (Wagner et al. 2002: 271). Biotechnology then forms part of a wider crisis whose underlying dynamic is the Weberian disenchantment of the world which I take up in the final chapter. 'Nature' and 'the natural' potentially become a rallying cry for what is regarded as valuable, genuine, healthy and human – against what is seen as industrial, rational, efficient, and profit-oriented. "The urgency of the appeal to nature", writes Macnaghten (2004: 548), "is itself a reflection of its contemporary "other", the threat of a technological runaway, possibly even "posthuman future" for which few if any of us [...] have adequate vocabularies or concepts'.

Transatlantic comparison, internal diversity, and the task of generalisation

All of the above concerns – nature as cosmological force, as placeholder for human virtues, as a sign of identity preservation and nostalgia, and as revolt against an alien 'post-human' world – are particularly relevant for European reactions to agbiotech. Such aspects of identity politics represent the cornerstone of my overarching argument. While it would be wrong to claim that moral and identity-related concerns about agbiotech do not exist in the US, they are typically less broad-based and less acute. Similar narratives about GMOs' 'unnaturalness' do materialise in US focus groups (Knight 2009) and US food marketing draws heavily on the positive connotations of naturalness (Sagoff 2004). Moreover, some of the underlying meanings – such as authenticity, tradition, and unadulterated purity – are found on both sides of the Atlantic. However, there are clear differences when it comes to the scope and intensity of concerns about naturalness. For example, in a comparative study of six Western countries, Rozin et al. (2012: 450) found that the US was the only country where the most common free association of the word 'natural' related to the adjective 'healthy' and only then to biological entities such as fruit, vegetables, water, etc. This indicates the prominence of utilitarian and self-enhancing values.

Civilisational dispositions and cultural context suggest that the mere presence (or salience) of ideas and values related to naturalness does not necessarily imply a similar centrality in attitude formation. For the majority of US citizens, nature as a cosmological and identity-related force exists primarily at the wilderness margins of society, while food and agriculture are subject to relatively strong instrumentalist/utilitarian tendencies. GMOs enter a cultural context that is not as easily unsettled by technology's symbolic and material implications – except perhaps on the East coast where anti-GMO mobilisation has been strongest and where history and landscape, to some extent, resemble the European intermingling of nature and society.

Kniazeva (2005) has studied the perception of GM food by American consumers and her findings support the idea that GMOs are potentially compatible with many core American self-enhancing values – such as personal improvement, growth, independence, practicality, efficiency (Schultz and Zelezny 2003) – especially if GM foods can be linked to joyful symbolic meanings and positive emotions involved in the practices of consumerism. A comparable study is not available for European societies, but it is worth recalling that, as noted in Chapter 2, for Europeans the 'usefulness' of agbiotech is often based on a self-transcending (societal) rather than a mainly self-enhancing judgement. This tendency fits well with premises of virtue ethics which I summarised earlier. By contrast, what worries a sizeable number of Americans are the potential health risks and ecological consequences of agbiotech. Due to my focus on less fathomable underlying causes of anti-GMO sentiment, I de-emphasise such mainstream concerns about physical risks. This does not mean that these approaches are invalid or that physical risks are insignificant. A sense of a new age of uncertain, less predictable risks, as postulated by the school of 'post-normal' science, the impossibility of strictly objective risk assessment (Levidow and Carr 1997; Stirling 2003), and the utility of precaution have reached the circles of policy-makers and experts alike, even if mainly in Europe.

Nonetheless, it is less evident that such concerns are at the heart of Europeans' opposition to agbiotech, unless they are packaged in a 'cosmological' or identity-related response. When reformulated, these anxieties represent a combination of 'pre-normal' sacred science and 'post-normal' insights. In this way, the second sense of nature as spiritual force joins the third notion of nature as a material structure and evokes a 'post-normal' condition of natural science in which unpredictability and potentially catastrophic side effects feature prominently. The rediscovery of the limits of scientific knowledge and human ingenuity is then often expressed with a characteristic 'animistic twist': by 'tampering' with nature, human beings may be bringing about their own demise, for ultimately they can never fully control this superior force or escape its inevitable 'revenge' (Wagner et al. 2001: 86). The image of animistic nature, the intuition that 'nature knows best', and the accompanying knowledge about human frailty and hubris can be seen as 'part of the European cultural heritage' (ibid.: 92), fashioned over the ages by an intimate relationship (of both benefit and dependency) with the natural environment. It represents another face of a broader civilisational disposition founded on an interactive relationship between nature and humanity.

Taking stock of the public debate, particularly in Europe, yields a picture that might comfort the defenders of 'sound science' and technological modernity. Because the European public does not predominantly argue on 'post-normal' risk-based grounds and instead mobilises a variety of images of nature, the charge of committing an irrational, 'naturalistic fallacy' is not wholly unfounded. It does, however, miss the point about the origins of popular resistance. When the EFB accepts that, in terms of modification and manipulation of plants or bacteria, '[i]n any useful sense modern biotechnology does involve a significant departure from what has gone on before' (EFB 1999: 11), then the 'naturalistic fallacy' constitutes perhaps an understandable, defensive reaction to radical change in symbolic and material terms. Both the moral ordering of the world and a way of life are seen to be in jeopardy. In the US, on the other hand, agbiotech is more readily compatible with a long-established and relatively accepted process of modernisation, rationalisation, and technological innovation. Resistance is largely confined to counter-cultural forces within society and has only recently begun to capture the attention of the mainstream.

Another conceptual challenge for comparative transatlantic analysis concerns Europe's internal diversity. So far, most studies have either concentrated on particular countries (e.g. Germany and Britain (Jasanoff 2005)) or examined the workings of scientific committees and political decision-makers at national and supranational levels (Toke 2004; Ramjoué 2007). An argument based on cultural and environmental history will sometimes find it difficult to generate plausible generalisations for the whole of Europe. The US has its own internal diversity, which will be further explored in Chapter 4, but it is arguably not on a par with European disparities. My way of dealing with this challenge is to posit the underlying similarity of the 'ontological condition' that characterises the relationship of European peoples' to nature. An interactive relationship with the land over centuries has given nature a special place within the cultural order, fashioning an elemental civilisational disposition towards nature. Popular resistance to GMOs draws on somewhat divergent discourses and values among European societies, yet to a significant extent, these are still unified by similar patterns of reaction against a perceived undermining of (often national) cultural identities, food cultures, cultural landscapes, and ideals of nature. Agbiotech is typically not regarded as a 'normal' technological development. Rather, it is seen as 'against nature', anti-cultural, a manifestation of capitalist globalisation or Americanisation, or as the final wave in the industrialisation of the countryside. Identity politics – connected to historical modes of living, feelings of belonging, and 'cultures of nature' – is the central component of effective popular mobilisation.

In countries such as the Netherlands or Spain, which remain relatively pro-agbiotech, the weakness of relevant identity aspects might account for the public's inaction and governments' regulatory lenience. By contrast, the successes of social movements operating at the national and increasingly the European level reflect the strong cultural flavour of political campaigning. These movements represent, on the one hand 'a traditionalistic reaction against modernity', but can equally be seen as being embroiled in the contest of a dominant against a repressed type of modernity. They are attempting to create 'an alternative modernity' (Eder 1996: 141). In Chapter 6, I describe this mobilisation as a neo-Romantic enterprise of 're-enchanting' an ever more disenchanted world. This popular movement is not primarily concerned with conventional political issues such as 'freedom, equality or liberation' but with '"irrational" issues [...and...] goods that can neither be divided nor distributed' (ibid.: 142).

Other scholars have also played down the significance of national and local specificities and have supported the use of Europe-wide generalisations. Wagner et al. (2001: 91) thus insist that

> [w]hen it comes to the basic questions pertaining to nature and life, or fear of global technology and economic developments, the European public implicitly demonstrate shared cultural roots. These roots transcend national boundaries, language barriers and north–south contrasts.

Marris et al. (2001: 72) equally stress that 'underlying socio-cultural factors' show genuine commonalities across EU member states. According to Wagner et al. (2001: 91), any observed differences, both in timing and in discourses of mobilisation, can be safely overlooked: 'these are located on a quite superficial level well above the substratum of Greco-Roman and Judeo-Christian undercurrents, which shapes the majority of European mentalities.' Although this fits well with my assumption of shared civilisational dispositions in Europe, it seems like a rather sweeping judgement. I will therefore return to the theme of intra-European divergence later in this book. Even subtle differences in themes of 'public outrage' are important, as they provide linkages of general environmental history with more specific cultural contexts and instances of identity politics.

Conclusion

The purpose of this chapter was to show that the concept of culture can be made analytically productive without dissolving it in constructivist narratives or fusing it with hegemonic structures of power. The cultural-political approach proposed here regards cultural differences and identities as socially constructed, but emphasises their historical endurance and (indirect) explanatory significance. The concept of cultural materialism, meanwhile, allows for the consideration of both materialist and ideational components of historical processes. From this broad conceptual agenda emerge the theoretical constructs of civilisational disposition and cultural identity, the former referring to deep-lying cultural outlooks or intuitions, and the latter to more overtly constructed worldviews and ways of life. Culture and civilisation do not so much undermine a *political* analysis as complement it by delivering a better understanding of the context within which political struggles take place. Context not only consists of the sediments of prior political decisions but also encompasses pre-existing cultural values, identities, and attitudes. Political discourses must resonate with these cultural 'resources' in order to appeal to the public at large.

Cultural features with the greatest relevance for the politics of agbiotech are public understandings of nature and objections to the perceived unnaturalness of GM foods and crops. These views fundamentally differ from the perspectives of most scientific experts and professional ethicists. To some extent, the cultural heritage of European societies stands in the way of a technology which appears to revolutionise the humanity–nature relationship by turning the former into a malleable resource for human purposes and by undermining a host of partly unconscious beliefs staked on the existence of a 'natural order'. In defence of traditional cultural identities shaped by a millennia-old ontological condition of interactivity with nature, many European citizens express an intuitive moral judgement on agbiotech, whereas this response is less acute in the US. The following chapter will flesh out these differences by exploring the political patterns of opposition to GMOs on both sides of the Atlantic. A culturally motivated rejection of unnaturalness is clearly not the only important element, but it pervades European public opinion and acts as the 'epicentre' for promoting greater regulatory stringency.

4
Cultural Politics and Resistance to GMOs

Introduction

In the previous chapter, I proposed a cultural approach and outlined how it could be partially reconciled with the study of politics. Important cultural elements, such as moral judgements and deeply embedded attitudes towards 'nature', could thus be joined by more observable political dynamics, namely the activities of interest groups, the impact of the media, and the dynamics of political systems and bureaucracies. To varying degrees, these are examined in this chapter. However, these factors cannot be understood outside a broader context which has deep historical roots and is constituted by pre-existing cultural dispositions. Thus, I explore the *cultural* politics of agbiotech on both sides of the Atlantic.

As I argued earlier (Chapters 1–2), in narrow causal terms, public opinion and successful NGO mobilisation seem to account best for the transatlantic regulatory divide. However, most existing explanations do not seek out the structural, historical sources of public unease and the often unconscious, value-based process of attitude formation. Qualitative studies offer crucial insights that complement the panoply of quantitative works on public opinion. But to establish a link to regulatory outcomes, one still needs to investigate organised opposition to agbiotech. Because the protagonists of the debate – civil society groups, NGOs, biotech firms – have to pay close attention to political as well as cultural opportunities, their actions, narratives, and achievements may, to a significant extent, reflect the broader cultural context. However, since cultural contexts never have a wholly deterministic quality and since other variables, such as political opportunities or the media, also play a role, my emphasis on the cultural dimension does not obviate the usefulness of a multi-causal analysis.

Furthermore, it is important to rethink the relationship between NGO activism and public opinion. Many accounts of resistance to agbiotech tend to ascribe its success to well-funded NGO campaigns that have targeted anxious consumers still reeling from recent food crises. While this is true to some degree, NGOs can also be regarded as politico-cultural 'entrepreneurs' who seek to capitalise on socio-cultural trends and powerful narratives (Bernauer 2003: 69). Popular unease alone is usually not sufficient for generating lasting resistance and NGOs act as mobilisers through discursive interventions, direct action, and political lobbying.

In the following, I examine the American and European contexts. Rather than trying to standardise the comparative output, I concentrate on those aspects that matter most in their respective political and cultural environments. Apart from the centrality of NGOs in both cases, in the US these core facets include the media, the dominance of scientific 'truth', the ethics of utilitarianism, the power of industry actors, and recent promising campaigns for mandatory labelling of GMOs. In Europe, key aspects comprise the multi-level system of governance, the tactics of anti-GMO movements and the agbiotech industry, as well as variations in cultural politics in different member states. Together, these elements constitute a comprehensive explanation of divergent levels of mobilisation and political influence. The analysis will equally highlight the links to the deeper cultural-historical context and to moral reasons for rejecting GMOs.

Understanding resistance to GMOs in the US

Composition, strategy, and impact of the anti-GMO movement

The impact of the anti-GMO movement in the US has so far been less decisive than in the EU. In contrast to the broad and vociferous antagonism aroused by plans to open parts of Alaska's Arctic National Wildlife Refuge for oil drilling,[1] the US movement has been compelled to rely on a narrower alliance of actors which – until the US media was galvanised by events in Europe – had greater success with lawsuits than with fomenting public outrage. The movement had also been unable to sign up 'influential opinion leaders' who would publicly oppose biotechnology (Hallman 2000: 6). Furthermore, the two-party political system has systematically frustrated anti-GMO endeavours by filtering out sceptical voices of the kind that have risen to prominence in the 'unsettled pluralism' characteristic of European politics (Gaskell et al. 2002: 373).

In the first two decades after the 1975 Asilomar Conference, anti-GMO activities were spearheaded by Jeremy Rifkin's Foundation on Economic Trends (FOET). By publishing several books and launching a number of lawsuits against federal regulators, Rifkin positioned himself at the forefront of a 'loose coalition of environmentalists, postmodern intellectuals and fundamentalist Christians' whose main arguments were grounded in variants of spirituality and strong ethical concerns about biotechnology's social and ecological ramifications (ibid.: 362). Without Rifkin's legal challenges, the 1986 'Coordinated Framework for Regulation of Biotechnology' would not have been created in this form or with such speed. 'Consensus workshops' were set up to explore the options for more widely acceptable regulations and the National Agricultural Biotechnology Council, created in 1989, was tasked with promoting regular and open discussions among the different stakeholders.

In the mid-1990s, it appeared as if agbiotech had become sufficiently institutionalised to expect a withering of the remaining opposition. But the political storm over GMOs across the Atlantic and the Monarch butterfly (1999) and StarLink (2000) episodes also reinvigorated the anti-GMO movement in the US, encouraging it to wage a major campaign from 1998 to 2003. Buoyed by generous funding from several foundations, dozens of NGOs[2] began to cooperate under the banner of the 'Turning Point Coalition' (TPC) whose activities culminated in 25 full-page advertisements in the *New York Times* over a period of 12 months, costing around 1–2 million US dollars (Reisner 2001). Supplementary actions and a considerable increase in critical press coverage accompanied the advertisement strategy. A successful defence of official organic food standards in roughly the same period further energised the movement.

To explain this surge in interest, it is necessary to view NGOs as partly strategic actors. Especially in the US, where financial backing and organisational acumen are often extremely important, public interest groups tend to carefully define their political objectives. As Reisner (2001: 1390) expounds, 'movements will adopt issues that they see as being directly relevant to their cause and ignore issues that are relatively distant or involve considerable [...narrative] realignment'. It is also essential to assess the status and capability of the movement as a political actor. From a rational-choice perspective preoccupied with the 'logic of collective action' (Olson 1971) – whereby diffuse (public) interests are difficult to represent effectively – the movement requires both a unifying narrative and an issue definition with potential for genuine public outrage (Bernauer 2003; Trumbull 2012).

Ethical concerns served as an adequate basis for Rifkin's flexible coalition, but they largely failed in terms of broader popular resonance. In a second phase of campaigning (1998–2003), the TPC's members emphasised a number of different issues related to agbiotech innovations, but they all used a narrative of health risks connected to the unknown consequences of eating GM food. While food and agriculture groups stressed the message of sustainable farming, environmental groups spread warnings about unintended environmental effects, and anti-corporate groups protested against monopoly control of the food chain, the central underlying theme was always utilitarian concern over health, environmental and economic impacts (Reisner 2001: 1402). Furthermore, each movement group adjusted their arguments to the overall aim of the TPC by blending their characteristic narrative with those of associated organisations. Anti-capitalist objectives and health concerns, for example, were integrated into the discourse of the environmentalist section.

Much of the impetus hinged on a clear, but fairly thin, shared storyline that called for much more stringent evaluation of health and environmental impacts as well as mandatory labelling (Toke 2004). Toke divides the anti-GMO camp into 'pragmatists' and 'activists' (or 'radicals') and highlights the fact that the bulk of the opposition was rather pragmatic-reformist in character. To name but a few examples, among the pragmatist organisations one would count the Consumers' Union, Public Citizen, the Earth Island Institute, the Union of Concerned Scientists, the Center for Food Safety, the Sierra Club, the Environmental Defense Fund, and the Natural Resource Defense Council. Among the more radical NGOs were Mothers for Natural Law, the Organic Consumers' Association, the National Family Farm Coalition, Earth First!, Greenpeace, Friends of the Earth, and the Council for Responsible Genetics, and more recently groups such as Food Democracy Now and the Alliance for Natural Health.

The positions of moderate NGOs were almost indistinguishable from mainstream European views. In the publications of the Sierra Club, for example, pride of place was given to the scientific case against GM crops and its uncertain health and environmental risks, with an occasional reference to equity considerations and issues of global justice. The overall stance is best summed up by the following statement: 'We urge the development of adequate regulatory, legislative, and other controls and that these decisions be based on a reverence for nature and life, as well as socioeconomic equity' (Sierra Club 2001). But apart from outlining the natural-scientific and socio-economic case against GM agriculture, the Sierra Club has been rather coy about the moral dimension of this struggle, merely noting that shared 'ethical principles lead us to respect and protect the natural world'.[3] Among food and public health groups, the Consumers' Union and the Consumer Federation of America have advocated mandatory safety approval before any genetically engineered food is put on the market and, especially more recently, both have vigorously campaigned for mandatory labelling.

Activist groups – such as the Council for Responsible Genetics – have used stronger language and introduced broader critiques of the new biotechnologies that converge on a transformative vision for sustainable agriculture. Greenpeace USA similarly employed a wide range of perspectives and merged concerns about social, economic, ecological, and public health impacts, while the National Family Farm Coalition stressed socio-economic concerns and called for a strict liability and testing regime that would have to precede any arrangement of coexistence of GM crops with conventional or organic crops. The Organic Consumers' Association has also had considerable ethical and socio-economic doubts about agbiotech, prominently displayed in campaigns such as 'Millions against Monsanto', even if it emphasised familiar utilitarian concerns about scientific uncertainty, human health, and the environment.

Another important question concerns the scope and representative quality of the TPC. Some of its member organisations –particularly those from the environmental section – did not do much more than sign up and donate money to the common effort. Half-hearted opposition or even cautious support for agbiotech actually has a long history in the US environmental movement. Toke (2004: 16) observes 'a groundswell of moderate environmentalist opinion that was favourably disposed towards the new technology'. This positive attitude (e.g. see Elkington and Burke 1987), can clearly be discerned in environmental groups from the moderate or the conservative spectrum. The Center for Global Food Issues (CGFI) assembled a conservationist coalition in favour of 'high yield conservation' in 2002 and made the case for using agbiotech technologies to simultaneously solve the crisis of world hunger and protect the planet's endangered species and wilderness. The Nature Conservancy has also published a broadly favourable selection of viewpoints on agbiotech. The idea of GMOs as potential tools for more environmentally friendly food production has real currency with the more mainstream environmental protection organisations in the US, but they have to be careful not to alienate some of their members who may feel more strongly about the ethics of genetic modification.

Besides non-participation in the TPC, another measure of anti-GMO sentiment are NGO statements – or indeed the lack of strong policy statements – on agbiotech. Searching for positions on GMOs on the websites of the Wilderness Society, the League of Conservation Voters, or the American Wilderness Coalition proves to be a fruitless undertaking. One clear exception is strong opposition against authorising fast-growing GM salmon which might represent a threat to the genetic integrity of wild salmon if it were ever to escape from its inland production sites. The Audubon Society, on the other hand, is now presenting both favourable and critical opinions about GM crops. Finally, the National Consumers League concentrates on food safety aspects and merely insists on high standards of consumer protection with regard to new food technologies.[4]

Given the oligarchic structure of most US-based NGOs, the widespread relegation of members to 'checkbook' activists, and the proliferation of individually tailored marketing and funding campaigns (Brulle and Jenkins 2006), the link between official campaigns and grass-root sentiment is somewhat diluted. The idea of seeing a more accurate picture of public opinion through the mirror of NGO campaigns is therefore riddled with caveats. It is not entirely clear whether environmental organisations have adequately reflected the wishes of their members or if they have exceeded them, except for those NGOs (such as Friends of the Earth or the Sierra Club) which have functioning participatory procedures in place. But this problem seems less decisive if one returns to the notion of NGOs as politico-cultural entrepreneurs who have as much freedom to exploit political (and cultural) opportunities as politicians in representative democracies. Their strategic narratives and selective emphases therefore reveal some of the cultural dynamics of American society. Most NGOs' comparative – if gradually weakening – reluctance to take a firm position on GMOs is associated with an atmosphere of strategic caution. Arguably, this attitude will only shift 'when they see a substantial public outrage potential or when public aversion is already high' (Bernauer 2003: 91).

Overall, the 'Turning Point' campaign certainly helped to diminish political differences between the various member NGOs, but its overall effectiveness can be questioned. Numerous indicators drawn from social surveys and commentaries have marked out the period of 1998–2001 as the time of the greatest public scepticism over agbiotech. This correlates with the activities of the TPC, which petered out by 2003. It may be best to regard American NGOs as leaping onto a trend of criticism that was spilling over from Europe and was fomented by an increase in sceptical press coverage. One could equally ask why the TPC would have faded away if it had truly triggered the desired explosion of public outrage. A turning point had not yet been reached in the US debate, although the financial and political boost generated by a collective campaign has certainly shaped and sustained the ongoing efforts of the anti-GMO movement.

By the mid-2000s, as the prospects for a major regulatory sea-change were dimming, various US groups turned to more sectoral, piecemeal strategies. Two of these strategies are only briefly summarised here because they have already been described in Chapter 1. First, the Center for Food Safety launched a whole number of lawsuits against the planting of GM alfalfa, GM sugar beets, and GM eucalyptus trees. This tactic caused considerable delay and uncertainty, but was largely neutralised by the regulatory agencies' decision to conduct additional environmental impact assessments and expedite official 'deregulation' of the GM crops in question. A second strategy involved mobilising US farmers worried about the potential closure of major export markets for rice and wheat – due to potential refusal of GM crops or accidental contamination of conventional grain exports with unapproved GM varieties. This focus on the economic self-interest of major farming regions turned out to be highly effective. GM rice is still not being grown commercially in the US and, in 2004, Monsanto decided to write off its considerable investment in GM wheat (Schurman and Munro 2010: 144).

Third, another set of political-economic tactics involved consumer campaigns and threats of boycotting particular retailers and fast food chains. A small number of supermarket chains have been persuaded to phase out GM ingredients in their products (e.g. Trader Joe's, Wild Oats) or label all of their products to achieve full transparency (Whole Foods). McDonald's declared it would not use GM potatoes, Gerber Baby Foods removed all GM ingredients, the snack manufacturer Frito-Lay phased out ingredients made from GM corn and, in December 2013, General Mills decided not to use GM ingredients in its 'Original Cheerios' muesli brand. But overall, as outlined below, this section of the commodity chain has proved to be relatively resistant to activist efforts – albeit not to same degree when it comes to GM animals. The FDA's imminent decision on whether to approve fast-growing GM *AquAdvantage* Atlantic salmon[5] has already led several large grocery chains to declare that they would not stock GM salmon products.

Alongside attempts at pressuring large retailers and manufacturers, another market-oriented action involved the creation of a non-GMO label. Though not regularly tested, organic food was already required to be GM-free since 2000. The 'Non-GMO Project' has the ambition of providing a GMO-free supply chain for manufacturers, alert US consumers to the pervasive presence of GM ingredients, and ultimately reduce the demand for GM crops. Officially launched in 2009, the scheme has rapidly grown and, by September 2013, non-GMO verified products accounted for $3.5 billion of US food sales[6] (which is still well below 1% of total food sales). Nevertheless, this success has come with some drawbacks (Roff 2009). The potential commercial value of non-GMO certified products is attracting competing certification bodies, which might lead to a watering down of purity and verification standards. Crucially, perhaps, continuing expansion of voluntary non-GMO labelling might undermine the campaign for state-level and federal mandatory labelling by insinuating that organic and conventional agriculture can easily thrive alongside GM agriculture. But given that US regulatory agencies do not conduct routine verification and have not established official tolerance levels for contamination of non-GM products, the long-term coexistence of agricultural systems cannot be guaranteed (Hubbard and Hassanein 2013).

The three strategies just outlined were the main focus of anti-GMO activism after the end of the TPC in 2003. They have clearly had an impact on the fortunes of the US biotech industry, since they caused obstruction, delay, and uncertainty, thus raising the 'expense-to-revenue' ratio for future investments (Schurman and Munro 2010: 119). But the tactics always implied a risk of distracting from the struggle for stricter federal oversight and mandatory labelling. The renewal of activist fervour since 2010, however, suggests that fast-growing, market-oriented initiatives such as the non-GMO project may have reinvigorated rather than undermined the movement. By rehearsing well-known utilitarian concerns (about public health and environmental safety), embedding them in a narrative of sustainable agriculture (small farms, reduced power of corporations, healthier products, food quality vs. quantity, greener farming methods), and – crucially – emphasising the norms of 'consumer choice' and the 'right to know', the movement has begun to mount a powerful campaign for mandatory labelling. Spearheaded by new NGOs, such as Food Democracy Now, and campaigning networks, such as GMO Inside, 'Right to Know – GMO', and 'Just Label It!', this fourth strategy has seen activists promote a whole number of bills and referendums on labelling in over 40 of America's 50 states.

Compared to the resounding defeat of previous ballots (e.g. Oregon in 2002), more recent referendums were often lost by extremely narrow margins – for example, 49–51 in California (November 2012) and the same result in Washington (November 2013). Labelling bills in Maine (June 2012) and Connecticut (June 2013) have already been passed but will not be implemented until a special clause is triggered, which requires several neighbouring states to sign similar labelling laws. A labelling bill in Vermont was passed in May 2014 and will take effect in July 2016, unless the biotech industry's legal efforts manage to overturn it. If mandatory labelling in many US states or even at the federal level were achieved, this could prove to be a game changer for the anti-GMO movement because it would confront often oblivious American consumers with visible GM products on a daily basis.

But given the uncertain long-term impact of the recent surge in activism and legislative activity, which factors best explain the persistent permissiveness of US agbiotech regulation? Considering the potential responsiveness of the regulatory system to political concerns and interests – for example, through state-level or federal legislation or guidance and rule-making by the regulatory agencies themselves – institutionalist accounts of how the industry-friendly USDA became the lead agency, or of how 'regulatory cultures' became entrenched, do not provide complete answers on their own (Pollack and Shaffer 2009: 73). Rather, as I argued in Chapter 2, a combination of political economy and cultural politics offers important additional insights.

The structural and instrumental power of the 'Food Chain Network'

The enormous power of corporate actors in the US can be conceptualised in a variety of ways. Drawing on Clapp and Fuchs' (2009) typology, *structural* power refers to the central role of agbiotech corporations and food manufacturers in the US economy, including employment opportunities and tax revenues. This status has been further reinforced since the 1980s by a 'market-oriented', laissez-faire regulatory philosophy which regards new technologies as drivers of business success and national economic competitiveness. By 2011, the US accounted for well over half of the world's biotechnology firms and over 60% of total industry revenue (Edge 2011: 83). Agbiotech corporations have further increased their structural power through strategic investments and have achieved 'near-oligopolistic' control over the domestic supply of seeds for staple crops such as GM cotton and GM soybeans (Falkner 2008: 144). What may have started as a close partnership with farmers (Schurman and Munro 2010) may have gradually shifted towards relations of dependence.

Instrumental power – the political-institutional and financial preponderance of agbiotech proponents – is best captured by Toke's (2004: 133) analysis of policy networks. He concentrates on the 'Food Chain Network' (FCN) which convenes weekly meetings of biotechnology companies, agro-chemical companies, farmers, seed and grain distributors, food manufacturers, and food processors. Member organisations and companies agree on a number of fundamental principles and interests, such as the notion of 'substantial equivalence', and often lobby the US government on international trade and domestic policy-making. Their potential influence has been considerable: 'When in agreement, this network has been virtually unstoppable in translating its requirements into regulatory action' (ibid.). Not all farming organisations are part of the FCN or share its underlying faith in the industrialisation and vertical integration of agriculture. The National Family Farm Coalition (NFFC), which participated in the TPC, opposes both of these objectives as well as the current lack of mandatory labelling. The medium-sized American Corn Growers Association has also adopted an anti-GMO position.

Nevertheless, the largest and most powerful players, such as the American Farm Bureau Federation, are generally in agreement with the aims of the FCN, not least because they include many of the largest and most profitable agricultural producers (Bernauer 2003: 96). It is therefore possible to conclude that 'US agricultural biotech politics has been dominated by a strong and cohesive coalition of pro-biotech up- and downstream producers and farmers' (ibid.: 73). Whereas some consumers, farmers, and retailers may have, over time, become slightly more sceptical about GMOs, seed providers and food manufacturers (namely the Grocery Manufacturers Association) have largely remained staunch supporters of agbiotech: 'the top end of the U.S. commodity chain [...is] tightly sealed against activist influence' (Schurman and Munro 2010: 190). The values animating the 'Food Chain' policy network are productivist, Promethean, industrialist, and largely neo-liberal. Large-scale farmers and food manufacturers regard the agbiotech revolution as economic progress in a world understood in terms of utilitarian benefits and controllable risks. The socio-economic consequences of agricultural change are systematically subordinated to a master narrative of technological innovation.

But, of course, the food chain remains 'intensely politicised territory' (Herrick 2005: 291) where agri-business and lawmakers must keep a close eye on developments at the grassroots. 'Food Disparagement Laws', which make it easier for companies and farmers to sue critics for libellous comments, exist in over a dozen US states. Another method of pre-emption with regard to GM crops has become popular with state-level policy-makers. Following a number of local and regional referendums that produced moratoria on agbiotech in several Californian counties and labelling requirements on GM fish in Alaska, by 2007 lawmakers in at least 15 US states enacted bills 'pre-empting' local regulations on seeds and nursery stock in order to prevent the creation of local 'GMO free' zones (Edge 2011: 91). With a new wave of state-level labelling campaigns gathering pace since 2009, agbiotech supporters have tried to rely on their lobbying power and financial muscle to defeat most of the new labelling bills which I mentioned above. But since three of these bills have been passed and referendums have brought ever narrower victories against labelling, activist momentum increasingly amounts to a genuine threat which could ultimately produce a patchwork of state-level labelling laws.

Reports from early 2014 indicate that the industry coalition has adjusted its strategy and has begun pressing for federal legislative pre-emption. Thirty-seven industry and farming associations have banded together in the Coalition for Safe Affordable Food to lobby both the public and policy-makers for an industry-friendly federal bill. Their impact is not yet clear, but the FDA is already contemplating new guidelines on voluntary GM labelling at the national level. And if federal legislation were to be introduced to pre-empt (and thus invalidate) state-level labelling bills, it would probably also introduce a lax, voluntary federal labelling law which might enshrine the right of companies to use words like 'natural' to describe food products with GM ingredients. Furthermore, a new federal law would likely strengthen the safety reviews of new GM crops in order to ensure consumer confidence (Hopkinson and Bottemiller Evich 2014).

While anti-GM activists cannot outspend or 'out-lobby' the industry coalition, their biggest hope lies in fomenting the kind of public unease or even outrage which caused agbiotech to become marginalised in Europe. Discourses that resonate with many Americans and the strongly felt salience of 'consumer choice/sovereignty' are part of what Trumbull (2012) calls 'strength in numbers' – which can be contrasted with the assumption that wealthy, well-connected, and well-organised interest groups will typically prevail. Mandatory labelling would most probably boost the anti-GMO enterprise, but in order to obtain such legislation, the movement still needs to mobilise public opinion more than it has managed so far. Several obstacles will likely complicate these efforts.

The pacifying effect of media agenda-setting and scientific 'truth'

Returning to a theme already explored in Chapter 2, some scholars regard the role of mainstream media as potentially significant in shaping Americans' attitudes to agbiotech. Contradicting Shanahan et al.'s (2001) observation that media coverage after 1997 turned more critical of agbiotech, Priest (2000: 940) maintains that media reports have mostly been positive, 'perhaps even one-sidedly so'. Media-agenda setting theory and the notion of 'framing' offer a surreptitious way in which the media moulds the minds of American citizens. As outlined in Chapter 3, framing is an important element of social movement analysis and refers to the way in which events or conditions are described according to partial interpretations, thus shaping public perceptions. In the process, actors 'must tap into "sentiment pools" by way of mobilization of resonant symbols' (Crossley 2002: 135) – in this case, starvation in the Third World and technological progress as opposed to more complex notions of stewardship, sustainability, and justice. Furthermore, emotion and identity play a central role in determining the success of particular framings (Tarrow 1998).

As I argued in Chapter 2, there is limited evidence of media agenda-setting in the US. It is also important to note that, in principle, the instrument of framing is available to both camps in the struggle over public acceptance. Supporters of agbiotech have greater financial capacities and try to promote an uncritical approval of GMOs, while its detractors seek to foster a critical evaluation which requires a greater intellectual involvement and a willingness to confront official scientific assurances. Some commentators, such as Priest (2006: 197), view US mass media as less combative than their European counterparts and as making use of their agenda-setting power to 'restrict competing ideas while reinforcing a select number of viewpoints'. But Priest apportions most blame to a broader culture of scientific authority that delegitimises those who do not speak the same language or share mainstream objectives. In other words, the pre-eminence of 'sound science' extends well beyond the US regulatory arena into all spheres of life.

These considerations feed into the notion of the 'spiral of silence' which was first coined by Noelle-Neumann (1993) and is applied to the US debate over GMOs by Priest (2006: 196) in terms of a 'masking of dissent'. A spiral of silence generates an atmosphere in which it is more difficult for lay persons to disagree than to consent. Science as an 'arbiter of social truth' (ibid.: 213) has a long history in the US and, in a sense, parallels the judicial power of the Supreme Court. The latter's unassailable decisions – though deeply political – are meant to guide a multicultural citizenry composed of diverse faiths and professions. Yet, in practice, the ivory tower of scientific expertise tries (often successfully) to impose a rationalist-utilitarian order on American society which comes at the price of a 'delegitimization, perhaps even the repression, of nonutilitarian points of view' (ibid.). And at this point, a spiral-of-silence theory interested in social order and power relations can help to inform a cultural perspective on the public acceptance of agbiotech.

The supremacy of science and a preference for market-based solutions both shape the public response to GMOs more that opinion polls could possibly detect. In her 2006 study of US public opinion, Priest found moderate evidence for the applicability of spiral-of-silence theory. Her most intriguing discovery was that people are least likely to voice a dissenting opinion when it concerns the connection between morality and the environment, nature or religion (ibid.: 210). This can be interpreted to mean a number of things. First, some of the observed silence is surely a result of mere disinterest or lack of confidence, as a correlation with lower levels of self-reported knowledge appears to indicate. Second, it may also confirm the point that arguing from a socially delegitimised status requires a greater degree of courage, strengthening the idea of a spiral of silence. Third, however, it is instructive to look at the number of respondents associating GMOs with the category of morality. The number of religious references was about as large (115) as for risk–benefit arguments (126), but only 44 respondents talked of morality in terms that could be classified in the environment/nature category. If one were able to filter out the respondents who referred to GM crops rather than, say, GM animals or the abstract idea of modification, the number would probably shrink even further. This reassessment does not invalidate Priest's conclusion, but it suggests that the central argument of this book – the relative separation of 'nature' in the US from the moral evaluation of agbiotech – represents a plausible approach. By extension, the power-based, elite-led perspective on US public acceptance should not be seen as a direct contender to a cultural-political approach. It may instead be a complement that demonstrates how structures of political and cultural power reproduce and stabilise already dominant cultural meanings.

The pre-eminence of utilitarian values

Thus, the relatively accepting attitude of most Americans towards GMOs and the relative dearth of moral discourse have two major structural explanations. The socio-political explanation – media agenda-setting and a 'spiral of silence' – underlines the obstacles placed in the way of increased public awareness and critical reasoning. There are few effective channels to voice scepticism and the organised protest movement has until recently been based on 'entrepreneurial' interest groups rather than on a groundswell of public discontent. By contrast, a cultural explanation is more interested in the underlying, historically fashioned 'sediments' of people's attitudes and emphasises that even the policies enacted by political and economic elites are infused with cultural values and attitudes. Elites – such as lawmakers, regulators, and interest groups – are also burdened with the shadow of history as well as guided by their own conscious and unconscious beliefs.

The overarching question is whether anti-GMO campaigns effectively challenge established regulatory routines and values and succeed in bringing about a more restrictive handling of GMOs based on more precautionary procedures, including labelling, traceability, and post-release monitoring. The common purpose shared by many NGOs united in the TPC was a paradigm shift from industrial towards 'sustainable' agriculture. The latter denotes less-intensive farming and higher quality products while respecting animal welfare and protecting rural employment against excessive reliance on machinery. Such an alternative political and moral vision of 'Aristotelian agrarianism' (Thompson 2001: 225) has not always been explicitly promoted in anti-GMO campaigns for fear of alienating potential supporters. But without such an alternative paradigm it may become difficult to roll back agbiotech's takeover of US agriculture, unless the new technology is subject to major and well-publicised regulatory failures with serious consequences. An over-reliance on utilitarian arguments might therefore prove self-defeating in the longer term (Buttel 2005). As Thompson (2001: 227) notes,

> [t]he dilemma of sustainable agriculture in North America is [...] how to marry consumers' self-interested concern with healthy diets to a philosophical vision of agriculture that holds that this form of self-interest is symptomatic of a hopelessly corrupted social framework, one that has little chance of righting itself so long as people continue to order their lives according to norms of preference satisfaction.

The cultural neglect of the farmed countryside and the veneration of wilderness, detailed in subsequent chapters, is a structural feature of American society. The silence on GMOs by some of America's oldest environmental organisations and the support given to a countervailing vision of 'high-yield conservation' indicate that the idea of agriculture as a fusion of nature and culture is also rejected by an important part of the country's civil society. The utilitarian concerns that most animate Americans – health, diet, nutrition, and a clean environment – could, at least to some extent, be achieved through industrial agriculture. In this vein, Thompson (ibid.) claims that there is a danger that 'the sustainable agriculture community will place its faith in technology (albeit alternative technology), and [that…] philosophical questions will be ignored or minimized. That is, in many respects, the American way'. Agbiotech would presumably play a big part in the necessary increase in environmental control and bio-physical efficiency.

While this assessment seems especially valid for earlier phases of anti-GMO mobilisation, the movement has increasingly converged on the alternative vision of 'sustainable agriculture' – reflected, for example, in the rapid rise to prominence of the 650,000 member strong NGO 'Food Democracy Now'. If well-known opinion leaders were missing earlier on (Hallman 2000), prominent individuals such as Michael Pollan (popular food writer) and Marion Nestle (professor of food studies and public health) have become effective spokespeople for mandatory labelling, greater regulatory stringency, and agricultural reform. Thus, campaigners are hoping that the anti-GMO message will continue to gather strength and that a breakthrough might be achieved this time – either through more stringent federal legislation, a proliferation of state-level bills, or widespread consumer action leading to pressure from food retailers to segregate and label GM crops.

Ultimately, however, the movement will only succeed if it keeps mobilising public opinion and wins over undecided or disinterested Americans to the cause of sustainable agriculture. To do this, as outlined in Chapter 3, the cultural resonance of its core discourses is of utmost importance. As theorised by Benford and Snow (2000), the values evoked by those discourses should be not only *central* (i.e. essential) to people's sense of personal or national identity but also *salient* enough to outweigh competing values or political goals. The anti-GMO movement appears to have moved on from a reliance on utilitarian discourses (human health, environment, economic risks). These are clearly both central and salient concerns, but they suffer from a number of shortcomings that I described earlier. In the current push for mandatory labelling, anti-GMO discourses often emphasise the 'right to know' and 'consumer choice' (Smythe 2009). These are based on fundamental norms of transparency and individual freedom and can be expected to resonate strongly. America's cultural heritage of liberal individualist 'sovereignty' is difficult to reconcile with the paternalist flavour of the opponents of mandatory labelling. The latter's case hinges on the argument that if an 'irrational' public cannot be educated sufficiently to follow a rational risk/cost–benefit analysis, then it should defer to an enlightened elite of scientists and regulators. It remains to be seen whether the ideal of 'consumer sovereignty' is salient enough to overcome fears of potentially higher food prices (from labelling and crop segregation).

Truly resonating discourses can certainly boost the anti-GMO movement's 'cultural power' (O'Mahony and Skillington 1999). A small but significant victory was the agbiotech industry's recent acknowledgement that 'genetic modification', rather than the more technical and upbeat term 'genetic engineering', had become prevalent in American society.[7] Another more hyperbolic term currently en vogue with the anti-GMO movement is 'Agent Orange corn'. It is used to stigmatise Dow AgroSciences' 'Enlist' GM corn which is immune to $2,4-D$, a compound that was also an ingredient in the toxic 'Agent Orange' chemicals used to defoliate swathes of forests during the Vietnam War. This represents a clear rhetorical attempt at linking GMOs and industrial agriculture more generally with dangerous military blunders and harmful long-term consequences. Opposition to the military–industrial complex is implicitly equated to the struggle against 'Big Food' and its promotion of potentially harmful agbiotech.

Nevertheless, it is too early to tell if the anti-GMO movement's cultural power now overshadows the industry's undoubted political influence. Mandatory labelling, if it were to be achieved, might even turn out to be a Pyrrhic victory in the long struggle for sustainable agriculture. This is not merely because the US agricultural system is controlled by powerful lobbies and provides barely any funding for alternative farming methods. Family farms have become rare and, overall, only around 1% of Americans are employed in the agricultural sector (Morgan et al. 2006: 37). Given the still 'countercultural' character of campaigns for sustainable agriculture, it is likely that, for the majority of Americans, food and agriculture still resonate most with utilitarian discourses and values. With the exception of certain regions – especially on the East coast – most American farmland is sharply demarcated from the surrounding industrial estates, suburban housing developments, and spaces of 'wilderness'. Neither regulators nor the public perceive strong non-economic 'positive externalities' from agriculture. Government funding for agro-ecological schemes is designed to lower the environmental impact of agriculture without mandating a particular method or considering potential social or cultural benefits of farmed landscapes. If a broadly accepted public good is at stake, it is the production of low-cost food products for the nation. In stark contrast to Europe, when asked about rural amenities, most US respondents do not reveal a commitment to cultivated, 'cultural' landscapes (Baylis et al. 2008).

Within the current US cultural context and political system, perhaps the march of a new, powerful technology simply cannot be stopped. One of the stalwarts of the anti-GMO movement has already made a step in the direction of technology-reliant 'sustainable agriculture' and maybe others will follow him. Jeremy Rifkin, the early leader of the 'moral revolt' against the emerging biotechnologies, declared his support for marker-assisted selection (MAS). This technique is reminiscent of traditional breeding methods, but with the added benefit of allowing the breeder to identify 'appropriate plant partners at the gamete or seedling stage' (Rifkin 2006) by scanning the plants' genetic code for the desired genes. Rifkin claims that MAS offers all the advantages of genetic science without incurring unpredictable risks for the environment and human health (Stokstad 2006). Although he set out his moral opposition to biotechnology once more in *The Biotech Century* (1999), ethical concerns do not figure prominently in Rifkin's promotion of MAS. The 'species barrier' is preserved by the new genetic technique, but this advantage is now touted as a way of reducing risks to the environment and public health. The moral message of this move towards alternative technology is not a defence of the 'natural' but a social ethics of shared ownership and dispersal of political and productive power within society. MAS also won support from influential members of the mainstream opposition to GMOs in Europe, such as Stavros Dimas, European Commissioner for the Environment from 2004 to 2009. In Europe, however, new food or agricultural technologies must cope with a cultural context where agriculture and 'nature' are intimately connected, as subsequent chapters will demonstrate.

Understanding resistance to GMOs in the EU

The European politics of agbiotech needs its own narratives and categories of explanation, not least because the widespread interpretation of European scepticism as based on 'irrational fears' stoked by a series of food crises is not wholly persuasive. Given the drawbacks of a mainly event-based explanation of public outrage, which I outlined in Chapter 2, the significance of the BSE crisis should instead be construed as a 'window of opportunity' (Ansell et al. 2006: 119) which reflected a particularly dramatic failure of regulatory institutions and policy-makers and thus galvanised civil society organisations, the media, and the public at large. Anti-GMO movements in Europe proved adept at exploiting political opportunity structures, turning agro-food commodity chains against agbiotech, and tapping into the moral sentiments, cultural values, and identities of European societies.

The EU's political structure: Why the Union listened to the people

While political economy perspectives and cultural–political analysis offer indispensable insights into the success of European anti-GMO mobilisation, institutionalist approaches and bureaucratic politics – which I reviewed in Chapter 2 – significantly contribute to our understanding of the process of regulatory 'ratcheting-up' (Bernauer 2003). Bernauer and Meins (2003) thus highlight the link between NGO campaigns and the EU's multi-level regulatory system which offers numerous channels for access. Beginning this brief discussion with the European Commission, the political controversy over agbiotech put the EU's executive arm in a difficult position. Many EU bureaucrats have long seen biotechnology as another instrument for developing the EU into one of the world's most successful knowledge-based economies. A second principal motive for pursuing biotechnology-enabling policies is the Commission's commitment to a multilateral trading order founded on clear rules enforced by the WTO. Each additional month of regulatory dithering increased American discontent about the obstructive European framework. The drawn-out WTO dispute settlement procedure, as well as its eventual outcome, put the spotlight on the Commission's attempts to reconcile the disciplines of global trade with the reservations of numerous member states (Seifert 2006; Zurek 2007).

Furthermore, the EU's executive body has attempted to gradually move away from its overly technocratic and elitist image (Ansell 2006). Partly to justify further delegation of national powers to supranational institutions, the Commission has often professed to be acting in the name of European citizens, not least when polishing its international credentials on conflict resolution, development aid, or environmental issues (Bretherton and Vogler 2006). Its 'mediative regulatory policy style' (Skogstad 2006), characterised by the search for compromise and avoidance of direct confrontation with member states, has been accompanied by a growing appreciation of the importance of democratic legitimacy. Yet, the attraction of celebrating democratic norms and defending political control over new technologies is counteracted by the Commission's technocratic instincts: its preference for 'hard' scientific facts, the strengthening of scientific committees, and the creation of EFSA as a response to regulatory crisis over agbiotech (Torgersen et al. 2002: 70). This ambivalence has not helped to restore public trust in regulatory institutions, with many citizens complaining about institutional complexity and a 'sense of a regulatory vacuum' (Marris et al. 2001: 63). For them, the scientific-technical approach to GMOs is a key reason for feeling ignored by overlapping regulatory authorities.

A more direct challenge to the Commission's tactics continues to emanate from various member states. The Commission obtained a significant harmonisation of biotechnology policy in 1990, but has found it increasingly difficult to make use of its gradually expanded authority (Toke 2004). Its mediative policy style was suspended for a period of time after 2005 when WTO pressure had reached a new high (Lieberman and Gray 2006). Even then, attempts to muster the Council of Ministers' authority to lift national bans on certain GM crops were regularly rebuffed. The more general problems posed by legislative harmonisation have been discussed in Chapter 1. The de facto moratorium on new authorisations eventually came to an end due to the phenomenon of 'ratcheting-up', whereby EU-level regulation typically reflected a compromise that was significantly closer to the preferences of GMO-sceptical member states. This happened because a strong incentive, which has always motivated many new European directives, is the preservation or improved functioning of the European Single Market. With regulatory fragmentation, the putative benefits of the common market – such as economies of scale for businesses, the free circulation of goods and greater price transparency for consumers – are in danger of diminishing.

The upward trend in legislative stringency is especially interesting because one could have equally expected a political deadlock due to intense disagreement. However, a number of factors – 'partial national autonomy combined with domestic pressure, preferences of the EU Commission, and the co-decision procedure [with the European Parliament]' – ultimately explain the occurrence of a trend towards 'ratcheting-up' (Bernauer 2003: 110). The safeguard clause, which was retained in subsequent revisions of the main EU directives on GMOs, allowed temporary bans on specific GM crops and thus ensured a partial autonomy of member states. Despite its overt misuse and reluctance to make the dissident case on the basis of widely accepted scientific findings, only France's safeguard ban on Monsanto's MON810 GM maize has so far been challenged in the European Court of Justice (ECJ). The court ruled against the French government, sparking a succession of further administrative and judicial proceedings at the national level which are ongoing at the time of writing.

One final institutional ingredient is the European Parliament. Many analysts have argued that the Parliament has been somewhat overlooked, although its role in the gradual strengthening of agbiotech regulation has been crucial (Toke 2004; Andrée 2005; Seifert 2006; Burns 2012). Directive 2001/18 makes it mandatory for Parliament to be consulted on each new authorisation of a GMO, and genuine power is conferred by the expansion of the co-decision procedure through subsequent EU treaty revisions. The Parliament has always been interested in enhancing its political profile and a pan-European controversy over agbiotech represented a genuine opportunity. Even more importantly, the assembly was beginning to build its reputation as the champion of a 'fictitious European public' (Torgersen et al. 2002: 70), and specifically on consumer and environmental issues. A good proportion of the 'institutional activists' recruited by the NGO-led anti-GMO movement are parliamentarians or their aides (Ansell et al. 2006: 115). Parliament's radical legislative amendments – typically rebutted by the Commission or the Council – have often fared much better a few years later. Rules on liability, compensation, or mandatory public consultation were initially proposed by the European Parliament and have by now been enshrined in EU legislation (Toke 2004: 180).

The above 'political opportunity structure' was successfully exploited by NGOs such as Greenpeace, Friends of the Earth, and Consumers International. These internationally-minded NGOs have a significant presence in Brussels and a network of contacts among Directorates-General, parliamentarians, and other important players. They are thus in a position to act as 'insider groups' within the EU and aspire to the status of 'agenda shapers' (Ansell et al. 2006: 115). Their targets encompassed the various EU institutions and large multinational companies with a strong economic presence in the Single Market. Yet another set of alliances could be constructed by cooperating with some more mainstream EU consumer and farming groups. Overall, this multi-pronged strategy – directed at multiple access points and targeting a variety of important regulatory players –contributed to the 'ratcheting-up' of agbiotech regulations. Nevertheless, it is doubtful that this regulatory trend would have occurred without public mobilisation and resistance by numerous member state governments (Tiberghien 2009). An important conclusion is, therefore, that the institutional environment merely offers a favourable political opportunity structure to express a deep-seated refusal which ultimately originates elsewhere.

The need for interpretive analysis

This qualification is crucial because an influential, essentially historical institutionalist account of agbiotech governance in Europe holds that a 'window of opportunity' (the mad cow crisis – BSE) and the EU's multilevel governance (with multiple veto points) constitute the core explanatory variables. NGO campaigns and targeted lobbying succeeded because they skilfully exploited this favourable political opportunity structure (Ansell et al. 2006: 99). Given my reliance on the notion of cultural context (with similar structural characteristics), it is indeed useful to consider the political and institutional context of agbiotech governance. But it is also justified to be wary of using political opportunity structures as an 'all-encompassing fudge factor for all the conditions and circumstances that form the context of collective action' – such as 'political institutions and culture, crises of various sorts, political alliances, and policy shifts' (Gamson and Meyer 1996: 275).

Regarding the 'window of opportunity', which is seen as a rare chance to revise an entrenched regulatory landscape, there are certainly indications that the 'watershed years' of 1996–1997 were heavily influenced by the breakdown of trust and 'moral panic' caused by a truly pan-European 'public-policy disaster' (Grabner et al. 2001: 15). There is little doubt that 'exogenous shocks' like the BSE crisis play a vital role in political mobilisation. When the British government admitted in 1996 that mad cow disease could spread to humans, public trust in expert advice suffered a severe blow, as the same experts had long insisted on the absence of such a link. Having based their policy-making on discredited information, governments were now eyed with equal suspicion because they were thought to listen to advice from industry-funded scientists who were more interested in economic gain than the protection of public health. The shockwave from the BSE scandal was further enhanced by a string of comparable events, such as the scandal over HIV-contaminated blood donations in France and food contamination with dioxins in various European countries (Gaudillière 2006). Many scholars thus argue that the lessons from these crises were crucial for the public perception of GMOs (Vogel 2003; Tsioumani 2004). NGOs, the media, and many citizens instinctively emphasised the connections between the BSE crisis and the prospect of GMOs entering the food chain and the environment. Another major triggering event, the 1996 arrival of Monsanto GM soya at a French port, was framed as a repetition of the BSE crisis by the French newspaper *Libération* which declared a nation-wide 'Mad Soya Alarm'. Such interpretations were seemingly corroborated in subsequent years by further events such as the 1998

Pusztai controversy[8] in Britain (Tsioumani 2004: 281). A sensationalist media seeking headline-grabbing stories made a vital contribution to this general upward trend in public saliency. The end result was an atmosphere of widespread 'public outrage'.

However, as I argued in Chapter 2, interpretive approaches – which are sensitive to perceptions, values, and the construction of meaning – allow us to understand why some 'events' (and not others) are perceived as threats or crises of some sort (Toke 2004: 99). For instance, neither the StarLink, nor the Monarch butterfly events in the US, nor the increasing incidence of 'superweeds' (resistant to one or several herbicides used for GM crops) has sparked comparable outrage and mobilisation. In Europe, despite repeated NGO campaigns, neither the regular detection of excessive pesticide residues in the soil and food products nor the increasing levels of endocrine-disrupting chemicals in freshwater have been elevated to the status of public policy disasters (Sato 2013). In other words, alongside the discursive skills and organisational resources of NGOs, the character and intensity of public attitudes is a key ingredient for deciphering both the significance of events and evaluating the resonance of NGOs' discourses.

Strategies of NGOs and corporations

Notwithstanding the importance of an interpretive approach, a pluralist explanation of EU regulatory evolution – based on interest groups and social movements – should be part of the analysis. Some insights from political economy perspectives have already been discussed in Chapter 2. But more remains to be said about industry organisations as well as NGOs. The latter have achieved a degree of influence not seen since the peak of anti-nuclear activism in the 1970 and 1980s (Bernauer 2003).

> They have cut down GM crops on test sites, pressured major food retailers to go GM free, demanded the applications of the precautionary principle in approving new GM crops, monitored nations and companies for compliance with the moratorium, staged media-savvy symbolic protests against the genetic patents, lobbied all levels of government in favour of a GM ban, and challenged the scientific claims of private industry and government agencies.
>
> (Ansell et al. 2006: 97–98)

The 'collective action capacity' of NGOs proved to be considerable because NGOs mustered a grand coalition of actors and crystallised their demands around issues such as labelling, traceability, a five-year moratorium, and highly stringent health and environmental precautions.[9] This pragmatic choice of core themes was accompanied by a high degree of flexibility, allowing NGOs to depart from dogmatic positions that many rank-and-file members may have favoured and instead focus on demands with greater chances of short-term political impact. Not only did different units of the anti-GMO movement manage to agree on the above demands, which were often considered politically feasible, but they also recognised the narrow boundaries that existing political opportunity structures afforded them. Policy-makers were generally receptive to societal concerns, but wary of undermining the privileged perspective of scientific expertise.

With this in mind, and understanding the media's propensity to 'latch onto "catastrophic risk"' (Seifert 2004), NGOs often decried the unknown long-term effects on the environment and on human health. As shown in previous chapters, moral and political critiques of agbiotech technologies may be foremost in the minds of many campaigners and citizens, but they are more easily brushed aside as distractions from the agenda of 'making biotechnology happen' (Jasanoff 1995; Hampel et al. 2006). Just like in the US, the fundamental objective of many European NGOs is to bring about a much more sustainable agricultural system, based on organic or small-scale farms, high-quality produce, and local, cultural notions of taste (Morgan et al. 2006). Yet, much like in the US, they have often been compelled to accept the dominant frame of 'riskocentrism' (Heller 2004) and appeal to principles of public health, environmental precaution, and commercial liability (in case of contamination) as a pragmatic political strategy to frustrate the march of agbiotech.

The crucial difference to US campaigns was that these efforts – which boosted the movement's regulatory credibility and political impact – were accompanied by more radical, culturally inflected discourses in its interactions with the broader public. Evocative rhetorical devices flourished, ranging from the UK's Prince Charles' defence of nature as the realm of God to criticism of 'aggressive' American capitalism (regarding Monsanto) and savvy combinations of environmental and moral discourse, such as 'pollution' and 'contamination'. These non-utilitarian, often openly moralistic discourses resonated with European publics and demonstrated the existence of a congenial cultural context. For example, in the UK, Greenpeace frequently used the language of 'genetic pollution' to condemn both the potential environmental effects and the 'moral transgression' symbolised by GMOs (Levidow 2000: 328), which is also implied by the notion of 'unnaturalness'. Because novel biotechnologies could be depicted as largely untested and unpredictable, the direct action network 'GenetiX Snowball' began to use a 'biohazard' symbol, reminiscent of nuclear danger, in its protests and publications. Friends of the Earth charged agbiotech companies with turning the revered British countryside into a 'genetic laboratory', thus evoking the story of Frankenstein (ibid.: 330–334). Newspapers from across the political spectrum recognised the potential resonance of discursive linkages with GMOs, and the term 'Frankenfoods' became a frequently used moral hyperbole.

Moral and political opprobrium was also a crucial discursive strategy in other European countries with high levels of opposition to GMOs. In France, the small-scale farmers union Confédération Paysanne painted agbiotech as a symbol of American capitalism and undesirable globalisation (Heller 2004). In Italy, similar discourses were employed by an anti-GMO coalition (Mobiltebio) which rallied a great variety of political actors – including over 500 smaller NGOs, small-scale farming organisations, and even right-wing political parties – in defence of traditional Italian production techniques (Ansell et al. 2006). A comparable, but even more broad-based movement emerged in Austria, where the popular tabloid *Neue Kronen Zeitung* spearheaded the campaign together with several environmental NGOs (Seifert 2009). In Hungary, conservative newspapers were the strongest critics of agbiotech and relentlessly linked opposition with key values of nationalist ideology, such as anti-American, anti-corporate, and anti-globalisation positions (Vicsek 2013). And in Slovenia, critics of agbiotech successfully drew on discourses of 'unnaturalness' and foreign provenance to undermine the new technology's public acceptance (Zajc and Erjavec 2012).

The role of agricultural traditions and food cultures as symbolic (and material) amplifiers of national cultural identities was a major catalytic factor. For instance, agbiotech's closeness to the practices of industrial agriculture provoked the ire of conservationists in Britain. 'This type of genetic modification,' English Nature, a major UK government agency, wrote in 1999, 'will make farming even more intensive and is undesirable in the British countryside where farming and wildlife must co-exist' (Lang and Heasman 2004: 181). Agbiotech was also depicted as a threat to artisanal farmers whose plight had long preoccupied the public's consciousness. Agricultural and food traditions are inextricably connected in some European regions, making it relatively easy for small farmers' organisations – such as Confédération Paysanne in France or Coldiretti in Italy – to build alliances with consumer and environmental groups.

Evoking the agro-food 'patrimony' of their nations, activists rhetorically linked GM foods and crops to the decline of family farms and the spread of low-quality fast food (Ansell et al. 2006; Kurzer and Cooper 2007). In France, a network of radical activists – the *faucheurs volontaires* – began to systematically uproot and destroy GM crops. These actions were also used in other countries such as Germany and the UK, but French activists justified them as acts of 'civil disobedience' (rather than covert resistance) in order to gain greater cultural and political legitimacy, drawing on notions of political commitment and sacrifice, and flaunting their struggle against the globalisation and industrialisation of France's agro-food culture (Hayes 2007). By 2004 already, 27 out of 48 French GMO research sites had been destroyed (McCauley 2011) and putting the activists' leaders on trial only further increased their popularity. Besides, the tactic of 'crop-trashing' did not only function as a highly symbolic direct action and as a putative 'public service' (Levidow 2000), but also made scientific research on GMOs, which often requires multiyear-long field trials, a costly and uncertain enterprise (Doherty and Hayes 2012).

A third condition of NGO success relates to their multi-level activity and their ability to 'read' the different layers of the European politics of food safety. Greenpeace deserves a special mention because its characteristic mixture of long-term campaigning and eye-catching direct action – coupled with a capacity for strategic adaptation to different political contexts – proved to be highly effective. Greenpeace's direct action campaigns often attracted favourable media coverage and its efforts (in November 1996) to block the delivery of Monsanto's GM soy to the port of Hamburg helped to create an important 'trigger event'. The organisation's centralised management structure enabled it to 'quickly roll out anti-GM activities across Europe' (Ansell et al. 2006: 106), and its permanent presence in Brussels gave it direct access to European policy-makers.

Similar to the US, much of European anti-GMO activism was aimed at fomenting public outrage and generating a credible threat of consumer boycotts. As outlined in Chapter 2, a highly competitive, relatively concentrated supermarket sector was acutely vulnerable to boycotts. In the UK, for instance, it only took the decision (in early 1998) of a small supermarket chain – Iceland Foods – to phase out GM ingredients from its own-brand products to spark off a 'chain reaction' which caused the whole sector to follow suit (Schurman and Munro 2010: 117ff.). In many other European countries, comparable dynamics could be observed, with large supermarket chain Carrefour playing the trend-setting role in

France (in late 1997). These boycott campaigns played out concurrently with rising political and regulatory contestation and greatly reinforced the arguments of those demanding strong, precautionary regulation and even a moratorium on authorisations of new GM crops – such as sceptical member states, DG Environment, and many European parliamentarians. To frustrate the import, development, and planting of GM crops in individual countries and bolster the coalition of GMO-sceptical member states in the Council, NGO campaigns also targeted domestic policy-makers. This was especially successful in France, Italy, and Denmark, among others, where environmental NGOs such as Friends of the Earth and Greenpeace built coalitions with national farming or consumer groups, often gathering widespread public support and exerting pressure on elected politicians. Political responses included the establishment of new food safety authorities (in the UK) or invitations for critical scientists to join regulatory committees. More dramatically, the Italian Environment Ministry announced restrictive new measures in March 1997 even before the oppositional movement had become sufficiently organised, and in Greece leading environmentalists were asked to join the relevant ministry (Hampel et al. 2006: 82ff.). Other governments managed to contain the anti-GMO fervour and preserve some room for manoeuvre by using various tactics, such as downplaying its importance and 'lying low' (Germany), engaging the public in elaborate consultation procedures (Denmark) or promoting the agbiotech agenda when the national climate turned out to be sufficiently tolerant (the Netherlands).

One of the questions arising from this anti-GMO surge is the ineffectiveness of countervailing strategies by industry lobbies and biotechnology enthusiasts. A major cleavage here relates to the divergent public appreciation of NGOs and biotech companies. Consumer and environmental groups generally command much greater legitimacy and trust because companies are primarily driven by the profit motive. Eurobarometer surveys from the early 2000s consistently revealed only a minority of citizens (30–40%) who were satisfied with industry objectives and in agreement with its general interests (Bonny 2003; Tsioumani 2004). Moreover, firms were largely wedded to traditional forms of communication based on simple narratives of 'progress' and the promises of 'new and improved' products (Hampel et al. 2006). Having failed to win over consumers, agbiotech companies found themselves at the receiving end of growing public antagonism. Once the industry took note of the emerging movement against GMOs, the tone of the debate had already been set. For instance, Monsanto's advertisement campaign[10] in June 1998 did not only come too late, but was also frequently regarded as dishonest and offensive (Ansell et al. 2006: 105).

To illustrate why Monsanto's strategy seemed parochial and short-sighted, it is worth briefly considering the literature on corporate globalisation. Globalisation theorists who emphasise the power of global economic integration to induce a convergence of business strategies argue that 'nationally based influences of culture and politics are often of only secondary importance in shaping and constraining the behavior of corporations' (Levy and Newell 2000). In the case of agbiotech, however, there should be a much stronger emphasis on the same authors' admission that 'social and cultural forces [... shape companies' ...] investment decisions, and the public stances they adopt toward environmental issues.' Clearly, the mantra of industrial policy-makers and corporations – 'making biotechnology happen' – was heavily influenced by the specific cultural and political contexts of the US and the EU. Prakash and Kollman (2003: 637) thus note that

> [n]ational policy agendas surrounding biotechnology regulation have been influenced by international forces but only to the extent that the domestic problem streams have made room for these issues and only in a manner dictated by domestic politics.

As I pointed out in Chapter 2 and elaborate on below, the EU's attempts at regulating agbiotech through conventional technocratic channels primarily failed due to public outrage and its impact on political institutions and commodity chains.

In addition, business strategies were shaped by the short-term path dependency created by prior regulatory decisions (Drezner 2007; Ramjoué 2007) which consolidated into very different GM production regimes on each side of the Atlantic (Harvey et al. 2002). In the mid-1990s, in the US, Calgene's marketing of *FlavrSavr* tomatoes was only halted by poor yields and Monsanto basked in favourable GMO adoption rates among American farmers, celebrating the 'most successful launch of any technology ever, including the plough'. Yet, during the same period, the British company Zeneca was treading much more softly (ibid.: 145–151). American companies could dispense with labelling and segregated supply chains and did not yet have to fear consumer reactions. Zeneca, on the other hand, carefully chose its product (American GM tomato puree) as the first entrant into the food market and contemplated growing GM tomatoes in Europe until the consumer backlash began in 1997–1998. Its home region had to be won over before it would consider further expansion. As Nigel Poole of Zeneca phrased it,

> Zeneca is a British company and we are fighting in this country. We have to defend ourselves to the London stock market. You can call it ethical, but it was also common sense. If you can't sell in your own country, you have really got problems... We saw a lot of problems. We always saw that Europe would be a lot harder than the US for selling GM foods. It was hard to see why, but it was to do with consumer instincts... We had to prove that we could do it. It was a real test for ourselves.
>
> (ibid.: 145)

Monsanto's bullish attitude and its ill-fated 1998 advertisement campaign did not please European companies who were acutely aware of consumer sensitivity. As the tide turned, it became evident that their caution had been justified and that tentative alliances with food retailers became unsustainable. European NGOs proved to be more adept 'politico-cultural entrepreneurs' than their industry opponents. In addition, the strategic, organised targeting of food retailers and processors by European NGOs drove a wedge between different producer segments, leading to a situation where biotech companies and input suppliers found themselves largely on their own. NGOs could suddenly count on the lobbying capacity of powerful retail chains which sought to reassure their customers with promises of no-tolerance policies towards GM ingredients (Hampel et al. 2006).

The collective action capacity and strategies of pro-biotech forces turned out to be inferior to the cultural and organisational strength of the anti-GMO movement. Assembling a diverse circle of actors, including biotech companies, seed suppliers, research universities, and industry umbrella groups, pro-GMO activity was reduced to advertising campaigns and conventional lobbying of government regulators who were by now beginning to feel the political pressure (Ansell et al. 2006). By contrast, the nascent pan-European anti-GMO movement not only received the backing of large parts of the European citizenry but was also accompanied by numerous local/regional campaigns, and a growing number of European regions seeking a de facto GMO-free status. Anti-GMO activism was literally closer to the people – culturally, politically, and geographically – which much enhanced its legitimacy, resonance, and influence.

Intra-European diversity and cultural context

Although there is a marked contrast between the impact of anti-GMO movements in most European countries and in the US, the differences among various European member states also need to be recognised. One major explanation posits that divergent agricultural systems can largely account for both transatlantic and intra-European differences (Kurzer and Cooper 2007). The US system is marked by economies of scale, focused on bulk commodities and export earnings, and constrained by consumer expectations of low prices (Gaskell et al. 2001). By contrast, over the last two decades the EU's Common Agricultural Policy (CAP) has gradually adopted a logic of redistribution and de-intensification. Historically, European farmers primarily produced for domestic or European consumption. The underlying economic structure of (on average) smaller farms and higher units costs, coupled with higher employment figures (5% of European population), have spurred a move towards higher value food products such as organic or specialist regional varieties (ibid.). However, the direction of agricultural reform is not purely economically motivated. Rather, it addresses the three planks of sustainable development: overproduction, social stability, and environmental protection.

The negative environmental externalities of agriculture are especially under scrutiny in northern Europe. More importantly, however, positive externalities – ranging from picturesque cultural landscapes to heritage/identity and biodiversity issues – feature prominently in European and national agricultural policies. It is commonly accepted that 'market failure' – such as not providing aesthetic stone fences, rare species of farm animals or wildlife, or causing rural depopulation and land abandonment – needs to be remedied by government intervention (Baylis et al. 2008). Farmers often receive direct payments for the upkeep of traditional land features or for specific agro-environmental services such as the preservation of agricultural biodiversity.

The characteristics of the European framework are best captured by the notion of 'multi-functional agriculture', an inclusive vision that addresses 'the economic, social, and environmental health of rural areas, or the sustainability of human life' (Richardson 2000: 78). Numerous stakeholder groups are supportive of this objective, but significant opposition remains, leading to an unsettled and often contradictory set of incentives. This also means there is no strong encouragement for technological innovations, such as GM crops, because the implied intensification of agricultural practices (higher yield, but potential environmental and socio-economic externalities) is not fully in tune with the general policy direction (Ramjoué 2007: 430). This evident tension may well give rise to accusations of economic protectionism (favouring conventional plant protection technologies and organic farming), but, as I argued in Chapter 2, there is scant evidence for strategic anti-GMO lobbying by industry and mainstream farmers' organisations.

Nevertheless, given that European agricultural policy still leaves significant leeway for national policy design, national agricultural priorities may well be correlated with governments' positions on agbiotech. Surveying EU member states' views on GMOs, a recent summary by the USDA's Foreign Agricultural Service (2013) distinguishes between

- *Adopters*: Member states which have cultivated Bt maize in 2013 (Spain, Portugal, Czech Republic, Slovakia, Romania) and those which might consider growing new GM crops with beneficial traits if the EU regulatory system allowed them to do so (Northern Belgium, Denmark, Estonia, Finland, Ireland, Lithuania, the Netherlands, Sweden, UK).
- *Conflicted*: Member states in which some groups (e.g. scientists, farmers, feed industry) are willing to grow GM crops, while others (e.g. NGOs, green parties, consumers) remain sceptical (Southern Belgium, Bulgaria, France, Germany, Poland).
- *Opposed*: Members states where most stakeholders and policy-makers are against GMOs (Austria, Croatia, Greece, Hungary, Italy, Latvia, Slovenia).

The report frequently mentions the economic interests of farmers and the feed industry and also links resistance to agbiotech with organic production and 'heritage' foods (sold under Protected Geographical Indications (PGI)).

Kurzer and Cooper's (2007) explanatory model draws on a more advanced political-economic logic. The authors theorise that a country will hesitate to support agbiotech or even reject the technology if it faces sustained domestic pressure from a 'green–green block' – an alliance of environmental and consumer groups with small farmers' organisations. The prime indicator for the emergence of such anti-GMO coalitions is the presence of a strong and deeply rooted, alternative food production system, most prominently in France and Italy.[11] Echoing the earlier discussion of resonating discourses, Kurzer and Cooper

(2007: 1037) argue that well-developed, alternative agro-food systems enable 'campaign strategists to frame risk in a concrete sociocultural context'. By framing agbiotech as a threat to regional/national food cultures and traditional or environmentally-friendly modes of farming, environmental NGOs were able to bring on board many other, usually less activist, groups and generate higher levels of public outrage. Indeed, in many European countries, campaign pioneers such as Greenpeace were not making much headway before joining forces with agricultural or consumer organisations (Hampel et al. 2006).

Kurzer and Cooper's data from the early 2000s shows mostly positive correlations of public opposition to agbiotech with indicators of alternative agro-food systems, the strength of domestic biotechnology industries, and governments' voting behaviour at the EU level. These connections are very suggestive, but a few 'outlier' cases complicate the picture. A northern outlier such as Finland – which has relatively tolerant public attitudes towards agbiotech, a fairly large organic sector, and usually supports new authorisations at the EU level – is justified by highlighting that 80% of arable land is devoted to pasture and arable fodder cropping. To some extent, this would shield most Finnish organic products (dairy and meat) from the potential threat of GMO contamination. But this tolerance began to fade when the EU authorised the cultivation of GM *Amflora* potatoes in March 2010. Soon after, the *Helsinki Times* (15 April 2010) reported concerns over genetic contamination and 'unnaturalness', and there are indications that environmental campaigners have begun to ally themselves with receptive politicians and food manufacturers (Real Food Partners 2013). Another prominent, southern European outlier is Spain which combines a relatively traditional food culture with growing acreage of GM maize, relatively high public acceptance of GMOs, and frequent government abstentions at the EU level. A small 'green–green block' led by environmental NGOs was formed, but has failed to curb farmers' increasing demand for GM maize or mobilise significant public outrage. Kurzer and Cooper (2007: 1049) astutely observe that 'late industrialization and late accession to the EU has produced a strong productivist, modernizing approach to agriculture' which is reflected by comparatively few PGI 'heritage' foods and a relatively low percentage of small farms.

While these are significant correlations, which are suggestive of causal processes, they should not be taken as conclusive proof for a materialist reading of anti-GMO mobilisation in Europe. A materialist interpretation would assume that the economic self-interest of those benefitting from alternative food production systems is the core explanation for divergent patterns of agbiotech policy-making – both within Europe and in transatlantic terms. While the relative strength of a 'green–green block' appears to largely determine the levels of public outrage and political pressure that can be generated, these are merely *proximate* causes. To elucidate the catalytic conditions (or potential *root* causes) of these patterns, one has to inquire into longer term historical processes and try to unearth how public attitudes and associated agro-food systems have come about in the first place. For instance, in agbiotech-tolerant Spain, discourses of tradition, alternative modernity, and unnaturalness may have failed to resonate more strongly because they may have been tainted by association with the era of the Franco regime (Sanz-Lafuente 2003). Modern, industrial agriculture, by contrast, may chime more easily with a widely felt desire for economic development, globalisation, and innovation. Similarly, in the agbiotech-friendly Netherlands, there is a relative lack of identity-related aspects of food and agriculture which could otherwise be activated discursively. For now at least, GMOs in the Netherlands remain inscribed into a narrative of economic progress, reinforced by the socio-economic imprint of a strong biotech sector and export-oriented agribusiness.

An interpretive, cultural–political explanation would therefore seek to uncover why and how the material and institutional structures of national agricultural systems were established. Such an account is, for example, offered by Sassatelli and Scott's (2001: 231ff.) analysis of Austria's agro-environmental reforms in the late 1990s. They describe a government bent on far-reaching public mobilisation but are mindful of emphasising the pre-existing or at times newly emerging values. Alpine farming, bedraggled by high cost structures and declining rural populations, needed the twin benefits of added value through organic farming and patriotic–traditionalist consumer loyalty to sustain itself. These aims became official government policy. Policy-makers often respond to existing cultural or political preferences, either because they feel compelled to do so for electoral reasons or because they broadly share the public's concerns (Zurek 2007). In the Austrian case, once the new agro-environmental trajectory had been chosen, the arrival of GMOs was greeted with barely veiled antagonism. Agbiotech's potential economic benefits might encourage unwanted agricultural intensification, while its cultural associations (standardisation and mass production) would undermine the appealing image of traditional and 'natural' Austrian products.

Yet, to deduce a primarily economic motivation from this reasoning would miss the point about the cultural–political origins of the agricultural trajectory. Throughout Europe, agriculture has not been treated like other declining sectors such as coal, ship-building, or textiles. The defence of agro-environmental reform is thus 'not only economically motivated in a strict sense [...;] rather, it stems from a political choice to keep a sector that is less productive than its transcontinental counterpart' (Gaskell et al. 2001: 103). Furthermore, traditional agricultural landscapes cannot be preserved in a self-contained or merely touristic fashion, unlike national parks for instance, because they rely on managed cultivation, human habitation, and rural employment opportunities. In other words, the 'jointness of production' (Johansen 2007: 9) with regard to economic and non-economic benefits renders agricultural policy a foremost domain of cultural politics.

Contextualising activism: Theorising public mobilisation

It may thus be premature to think of public opinion and policy-making as being mainly driven by the material interests of organised producer groups and environmental NGOs. Cultural context, appropriately chosen political agendas, and resonating discourses typically precede successful anti-GMO mobilisation. Returning to the earlier discussion of cultural resonance and power (Chapter 3), anti-GMO movements not only rely on political and economic opportunities, but also need to deploy their cultural power by appealing to widely held values and attitudes (O'Mahony and Skillington 1999).

The central point here is that European NGOs have only partly acted as instigators and 'creators' of opposition. More importantly, they have been 'interpreters' and 'transformers' of existing trends of rising concern over agbiotech, recognising deeply embedded, *central* values and labouring to render them *salient* enough for political campaigning. The decisive phase of NGO activity (1996–1999) and the sudden rise in critical media reports were a way of tapping into existing or emerging perceptions of agbiotech and thereby meeting an underlying cultural–political demand. The anti-GMO movement has 'amplified and reified social representations and turned them into salient issues' (Hampel et al. 2006: 88–89). This interpretation is largely congruent with data from opinion polls (see Chapter 2), which show a growing sense of public unease as early as the late 1970s, and strongly suggest that NGOs 'capitalised on existing, if vaguely articulated, opposition' (Gaskell et al. 2001: 112). As Bernauer (2003: 69–70) also observes, NGOs were most effective in 'piggy-backing' onto emerging trends and managed to turn public outrage into political leverage vis-à-vis market actors and regulators.

Some have questioned whether public mobilisation had to take on a highly visible form or whether the mere impression or probability of it was sufficient to push governments to act. Toke (2004: 43) favours the traditional model of mobilisation, whereby NGOs had to genuinely 'activate' consumers in order to kick-start measures such as retail boycotts. By contrast, Hampel et al. (2006) put forward a revamped notion of mobilisation which is tied to prevailing attitudes. What could be defined as 'latent' mobilisation does not require visible action. It is sufficient that decision-makers consider it likely that mobilisation will occur if they press on with controversial policies. The barrier for NGO influence is set even lower. To gain concessions from regulatory bodies, it may suffice that NGOs give voice to the public's grievances.[12] This theory of mobilisation contains a key insight: at the root of NGO activism, industry weakness, and government U-turns lies the public mood which, in many countries, is shaped by a strong and persistent concern over agbiotech's technological trajectory. The partially independent and 'constitutive' role of the public is significant because it counters the proposition that European citizens have been led astray by enterprising NGOs and a sensationalist media. The public's relationship with both civil society organisations and the media is complex. Lack of trust in regulators and a higher trust score for NGOs need not signify that Europeans follow the latter's sermons blindly while refusing to listen to regulatory authorities and politicians. Rather, as Lassen and Jamison (2001) have noted, 'lack of trust can be interpreted as an expression of the different foci of the public and the regulators as well as a cry for proper representation of "the public view" in the political process.'

NGOs may have become vectors of public trust because they seemed to be the only critical voice around, even if people did not always agree with their statements and actions. According to one major UK focus group report, NGOs 'were not a strong reference point in the participants' daily life and were not mentioned as a key source of information' (Marris et al. 2001: 63). They were widely seen as biased actors but credited with representing a counterbalance to other stakeholders whose interests were even more removed from people's concerns. According to focus group participants, the principal advantage of NGOs was that they broadened the debate to include at least some of more general societal and environmental issues. However, as the above analysis suggests, even European NGOs were only moderately successful in performing this function. Lassen and Jamison (2001) conclude that NGOs frequently did not properly convey the widespread, more fundamental, moral objections of the public because the 'physical risk' discourse also dominated in their dealings with regulators. In summary, therefore, while NGO activism and the relative weakness of industry responses was the proximate cause of widespread anti-GMO mobilisation in Europe, this could not have happened without a catalytic, cultural context in which critical discourses resonated more strongly than the promises of agricultural modernisation, economic efficiency, and functional foods.

Conclusion

While anti-GMO movements in the US and the EU have used similar tactics (political lobbying, consumer boycotts), European activists have also taken direct action (e.g. GM crop trashing). The different degrees of success are not principally explained by divergent levels of resources, variations in tactics, or even by political opportunity structures, such as party systems, institutional configurations, and regulatory processes. Nor is a simple metric of economic and political power sufficient to account for the varying influence of interest groups. Nor do disruptive events, such as various European food crises, and associated 'policy windows' provide a comprehensive explanation of sustained levels of public outrage in Europe. Undoubtedly, all these factors are important *proximate* causes of political mobilisation and regulatory change or stability. But they do not tell the whole story about why the public was galvanised by some events and not others, and why sustained public scepticism (and mobilisation) has hitherto only been achieved in Europe. Scholars who analyse political processes and opportunity structures often acknowledge the significance of a broader context for explaining the impact of anti-GMO movements (Schurman and Munro 2009). But they rarely seek to uncover the deeper reasons for why societal mobilisation succeeds or why public outrage can only be successfully stimulated by particular discourses.

The US anti-GMO movement prioritised a three-pronged emphasis on human health, environmental effects, and socio-economic aspects, which largely pertain to the realm of utilitarian thinking. Yet, NGO campaigns failed to provoke public outrage on a comparable scale to Europe because prevalent anti-GMO discourses have not been confirmed by major public health or environmental disasters, while pro-GMO discourses have touted the benefits of cheaper food and solutions to global hunger. The regulatory process systematically neglects socio-economic aspects of new technologies, and the supremacy of natural science pervades regulatory institutions, the legal system, and the public sphere. More radical discourses about the ethical ramifications of agbiotech largely failed to resonate with Americans. The combined effect of consumer boycotts and forceful political campaigning for the 'right to know' (i.e. mandatory GMO labelling) may well change the situation, although it remains unclear if there is sufficient public support for a paradigm shift away from GM agriculture. And the structural and instrumental power of the 'Food Chain Network' represents a formidable obstacle.

While the potential health and environmental risks of GMOs dominated their discussions with policy-makers, European anti-GMO movements also included threats to 'nature' and 'tradition' or dangers posed by 'genetic pollution' in their communication with the public and the media. In the EU, such ethical or moral objections are part of mainstream opposition and have achieved a high level of public resonance, forcing European agbiotech companies to accept stringent regulations and pursue cautious, incremental strategies of commercialisation. There remains considerable diversity among national governments' and their publics' attitudes towards agbiotech. Political-economic analysis – based on the relative power of interest groups and the presence of alternative agro-food systems – indicates recurrent correlations between anti-GMO sentiment and the strength of 'green–green' alliances of environmental NGOs and small farmers' groups (Kurzer and Cooper 2007). But an interpretive approach built on cultural–political and historical analysis is needed to identify the driving forces behind the emergence and persistence of alternative agro-food paradigms as well as the promotion of industrial, technology-based agro-food models.

In sum, the arguments presented in this chapter corroborate the centrality of public opinion and the strength of cultural and moral values alongside utilitarian (risk–benefit) considerations. Placing public sentiment at the core of the analysis, however, does not necessarily imply an event-based storyline based on an 'irrational' public response and on NGOs as powerful agents of media sensationalism. Neither NGO campaigns nor industry efforts were able to manufacture public opinion or instil radically new values through discursive politics. NGOs were crucial as communicators and amplifiers of public concerns, but their ability to set the agenda on their own is rather limited (Hampel et al. 2006). NGOs strongly contributed to the observed 'public outrage' by using resonating discourses and engaging in targeted lobbying and direct action – in other words, by drawing on pre-existing 'political and cultural opportunity structures' (O'Mahony and Skillington 1999: 112). To some extent, citizens delegated the task of critique to civil society organisations, even though their predominantly moral objections were not always raised publicly. NGOs found moral arguments to be less effective in a political context dominated by natural science and 'physical risks' (Seifert 2004), as well as long-standing agendas for technological innovation and economic competitiveness. As mobilising devices, however, moral and cultural narratives and images proved to be very powerful. The remaining two chapters will explore the cultural, historical reasons for this level of resonance. They will also discuss the more limited options for the deployment of such 'cultural power' in the US context.

5
Environmental History: Nature, Landscapes, and Identities

Introduction

An analysis of the contemporary politics of agbiotech is enhanced by taking account of the cultural context – that is, pre-existing, historically constituted values and identities. Following the approach presented in Chapter 3, culture is conceptualised as a middle way between essentialism and voluntarism, while historical evolution is understood in a dialectical sense, drawing on both materialist and idealist factors. A century ago, the French geographer Vidal de la Blache proposed a conceptual fusion by introducing the concept of 'milieu' 'which embraced not only the physical but also the cultural environment within which [...] judgements and choices are made' (Baker 2003: 73). The majority of scholars dealing with the nature–culture relationship (among them environmental historians, historical ecologists, and historical geographers) adopt a similarly integrated position.

In the first part of this chapter, I examine the deep-lying intersubjective understandings which I termed 'civilisational dispositions' vis-à-vis nature. These relate to an elemental structure of humanity–nature relationships, namely bifurcation in the US and interactivity in Europe. Bifurcation does not automatically mean hostility towards nature. It can also encourage veneration. It does, however, mean an 'externalisation' of nature and prevents it from playing a major role in everyday life or the 'middle landscape' which encompasses both agriculture and human settlements. The core argument focuses on long-term historical trends in humanity–nature relations. Following a succession of gradual cultural changes in American history, nature in the US is often represented by 'wilderness'. Several explanatory factors will be brought to bear, including aspects of religion, modernity, capitalism, nationalism, and the Western frontier. For European environmental history, an even longer historical perspective is chosen in order to underline its distinct cultural and economic trajectory through the ages. I not only examine medieval attitudes to nature, but also trace the way in which nature was transformed from a customary notion of moral order into a newer political-economic concept of resources.

The second part examines humanity–nature relations in the modern era. At first glance, Europe and America appeared to follow a parallel track of cultural evolution. Unrestrained exploitation of and hostility towards nature were gradually replaced with admiration and conservation. The idea of nature progressively turned into a rallying cry of Romantic philosophers and other critics who worried about 'losing their souls' amid the restless transformation wrought by industrialisation. Nonetheless, European particularities remain important: first, a less intense separation of the natural environment from human activities has developed; second, a major difference concerns the kind of 'nature' conjured up to defend the traditional lifeworld or promote alternative models of modernity. While American Romantics were able to realise a radical interpretation of nature through the wilderness landscapes of national parks, Europeans celebrated a more pastoral form of nature. I trace the development of these divergent 'cultures of nature' and argue that transatlantic histories reflect both specific material conditions and dominant ideas and visions.

Early modern rationalisation and increasing material exploitation are shared features of America and Europe, but this does not irrevocably fuse them into a common Western tradition. European societies also share some experiences with non-Western cultures, and the bifurcating tendencies of modernity have been implemented more successfully on the North American continent. Finally, Eder (1996) observes that modernity is riven by two movements interacting in complex, often antagonistic ways: they could be described as techno-scientific and aesthetic, pragmatic and identity-based, instrumental and non-instrumental. All the way through the modern age, this cycle of action and reaction – of landscape transformation and planning laws, of industrial food production and artisanal/organic products, of intensive livestock-rearing and animal liberation, of natural resources and sacred grounds – has persisted. This 'double process of intensification' (ibid.: 145) defines the cultural politics of nature – and with it the controversy over GMOs, an innovation that can be construed as the next step of instrumentalist control over the natural environment.

The *New Founde Land* of North America[1]

Following Columbus' 'discovery' of the Americas in 1492, the Spanish Empire established a foothold to the south of today's US, while the French settled in the region today known as Canada and followed the Mississippi down to the Gulf of Mexico. Sir Walter Raleigh set up the first English colony in 1585 in Newfoundland and this short-lived fortified settlement was to endure constant threat from Spanish Florida, the hostile climate, and indigenous tribes. Whatever the precise causes of eventual survival and flourishing, the future American nation was to be profoundly influenced by the cultural currents emanating from New England, the fervently Puritan region of the first settlements. These settlers had to adapt to their environment and cope with being displaced to a foreign soil. Stoll (1997: 56) captures their initial bewilderment particularly well: 'Europeans first contemplated the wild shores of America much like a painter gazes at a blank canvas. To them, America had no past, no history – it was a continent in future tense.' This feeling of emptiness was quickly superseded by older ingrained habits, traditions and practices. Stoll describes the settlers as planting their Gospel, their beliefs, and their intellectual traditions into the new soil, conjuring up the image of a European seedling that slowly grows into something recognisably independent. For at least two more centuries, 'the American continent was a seedtime of ideas and attitudes from European stock' (ibid.), either by tradition or continuing transmission from the Old World.

America as a Puritan utopia

The newcomers were not bent on modest adaptation to environmental conditions. Harsh winters and the lack of appropriate skills initially necessitated a more adaptive stance and an eagerness to learn from indigenous neighbours, but the mainly Puritan settlers were committed to a 'psychological and spiritual quest', in which '[a]ctual landscapes are less important than the landscape of the mind' (Slotkin 1973: 38–39). This reading foreshadows the courageous stubbornness of the Western expansion. The main point, however, relates to the culture of early American Puritanism and the deep unwillingness of most settlers to recognise anything other than religiously ordered hierarchies – the 'absolute sovereignty of God' (Bolt and Lee 1989: 80). This can be seen as an early move towards human exceptionalism which separates the human realm from nature.

Much of the Puritan worldview stemmed directly from studying the Bible. Until the discovery of America, Christian theologians had paid little attention to the kind of untamed nature that appeared to dominate the new continent (Stoll 1997). The European history of gradual environmental transformation was much more difficult to interpret as the clash of two elementary forces, of what was to become the binary opposition of wilderness and civilisation.[2] At heart, Puritanism was a modernising religious movement that rejected the ceremonial 'overload' of the Catholic Church, sought to restore the purity of the Gospel, and eliminate the pagan rudiments of medieval society. Hence, most Puritans were not predisposed to reproduce images of nature as benign bounty or pious solitude. Instead, their militant zeal and a culturally embedded vision of the ideal landscape as pastoral paradise helped to brand the nature they encountered as a 'howling wilderness'. This also legitimated the rapid transformation of wilderness into farmland. It is not clear whether the pioneers were, in fact, too close to wilderness to properly appreciate it, as claimed by Nash (1967). Certainly, trappers and farmers were unlikely to develop an aesthetic view of nature, but it does not necessarily follow that utilitarian considerations need to go hand in hand with deep-seated hostility towards the natural world. The integrated belief-systems of many indigenous peoples around the world represent powerful counter-examples.

Nash equally emphasises the importance of the Puritans' predominantly religious mindset. In this sense, Slotkin's (1973) 'landscape of the mind' originates in the Bible's Genesis and humankind's fall from grace. The paradise of Garden Eden is epitomised by a quintessentially pastoral landscape. The way in which the settlers imposed human dominion on the land, its animals, and later its indigenous inhabitants indicates that they were implementing their perceived Biblical destiny. Wilderness was to be transformed into a plentiful garden for the greater glory of God:

> Puritans believed in the millennial transformation of man and nature, the refuge in the wilderness, the duty to subdue nature and replenish the earth, the work of spreading the Gospel and raising a new church, the need to put religion before profits, the testing of the Puritan Israel in the American Sinai.
>
> (Stoll 1997: 70)

Naturally, there were alternative voices within Puritan society, moderating the image of a radical break with more established European traditions. Stoll (ibid.) presents the poet Anne Bradstreet as a conciliatory voice in the colony, arguing that her pastoral Puritanism was just as legitimate as the image of a 'howling wilderness'. The study of sources from the Renaissance and from classical antiquity led Bradstreet to a much more benevolent relationship with the natural world. Her writings could be construed as the expression of a minority view.

Liberated from centralised ecclesiastical authority, imbued with great religious zeal, and faced with years of hardship, Puritans paid closer attention to their natural surroundings than their European relatives. Possibly there are still some remnants of medieval, animistic nature in their stories. Whether wilderness is benign and beautiful or brutish and devilish, there is an almost personified deity residing within it. At the same time, there is a clear attempt at systematising and simplifying the relationship with nature. To forge a coherent religious destiny, nature needed to become fixed in its essence. It is at this point that nature and humankind begin to drift apart conceptually before doing so physically as well.

The environmental impact of early Puritan practices

By the late seventeenth century, the onslaught against wilderness was well underway and New England's landscape was fundamentally transformed. The abundance of trees encouraged the antithesis of conservation, visible in increasingly bigger houses made from timber (Beinart and Coates 1995). By European standards, American settlers completed in a matter of decades what had taken Europe's inhabitants many centuries. But regardless of how transformative the outcome, the subsistence-oriented practice of settler farming represented a continuation of European habits before the advent of capitalism. Merchant (1989: 149) thus writes that many settlers still accepted nature as an animate being in need of being 'wooed', which corroborates the assumption that medieval sensibilities were still influential. At a socio-economic level too, English traditions of collectively managing the commons remained strong, only to be gradually replaced by a desire for private ownership of property (Donahue 2004). Despite their millennial objectives, the settlers were culturally still rooted in European practices. To some extent, the old agricultural patterns persist on America's East coast even today. 'After farmers quit the thin stony soils of New England for richer western prairies', write Beinart and Coates (1995: 69), 'secondary forests soon recolonized and one often stumbles upon a tumbled-down dry stone wall when walking in the woods.'

In many ways, the American settlements were an offshoot from European civilisation and the novel trends that were beginning to be felt in Europe (e.g. individualism, capitalism) found a more receptive environment in the New World (Bradbury and Temperley 1989). Although cultural dependence on European ideas and fashions persisted for many centuries, sustained cultural cohesion would have required regular interaction in all spheres of life.

> Whatever the nature of the forces at work, it was plain that when people and their institutions crossed the Atlantic they suffered a sea-change, and that, whatever their plans, and however determined they might be to preserve European practices, they invariably ended up with something different.
>
> (ibid.: 5)

Much of the speed and intensity of the American transformation was due to the Western expansion which began in the late eighteenth century.

The winning of the West

The frontier experience

In July 1893, the historian Frederick Jackson Turner presented his now famous paper (*The Significance of the Frontier in American History*) to a special meeting of the American Historical Association. He argued that the frontier of westward expansion had closed and thereby concluded the first period of American history. In what has been dubbed 'an environmental explanation of Americanism' (Potter 1954: 22), Turner claimed that the defining characteristics of the American frontier were the availability of free land and the mode of living in an untamed environment. European settlers were inevitably caught up in a transformative process as they joined the westward-moving frontier of settlement.

> The frontier is the line of most rapid and effective Americanization. The wilderness masters the colonist. It finds him a European in dress, industries, tools, modes of travel, and thought. It takes him from the railroad car and puts him in the birch canoe. It strips off the garments of civilization and arrays him in the hunting shirt and the moccasin. [...] In short, at the frontier the environment is at first too strong for the man. He must accept the conditions which it furnishes, or perish, and so he fits himself into the Indian clearings and

> follows the Indian trails. Little by little he transforms the wilderness, but the outcome is not the old Europe, not simply the development of Germanic germs, any more than the first phenomenon was a case of reversion to the Germanic mark. The fact is that here is a new product that is American.[3]

However persuasive this nationalist stance was at the time, the 'Turner thesis' subsequently attracted numerous critiques. In the twentieth century, many scholars reproached Turner for systematically erasing considerations of gender, ethnicity, class, urbanisation, and industrialisation from his study. Furthermore, geographers, such as Sauer (1963), reworked Turner's determinist premises into more accommodating, 'possibilist' perspectives. They also located the frontier in a larger historical framework that recognised ecological and anthropological aspects of the American continent, recalling the fact that American history did not start with the arrival of European settlers.

In terms of timing, however, Turner proved to be remarkably prescient. The nineteenth century was the period during which Americans' view of nature underwent rapid changes (Stoll 1997: 100) – towards more binary attitudes. The civilisational disposition of Americans became more defined. In line with the currents of modernity, rationalisation, and Romanticism, American attitudes towards nature were increasingly marked by bifurcation rather than integration.

The persistent relevance of the frontier thesis

In contemporary writings, the frontier thesis often gets short thrift as a nationalist or Romantic act of myth-making (e.g. Cronon 1996: 76). Yet, apart from the values it promoted, some fundamental changes were genuinely taking place (Burchell and Gray 1989; Arnold 1996). At a purely environmental level, the transformation was breath-taking. The landscape where trappers, hunters, and native tribes had been the only human presence now encountered the agricultural might of large numbers of human immigrants. Traditional slash-and-burn practices were superseded. From a historical perspective, the speed and scale of environmental change has few historical parallels (Arnold 1996), even when considering the experience of other frontier societies in South Africa or Australia. As much as the frontier environment may have influenced its new inhabitants, it was also their own cultural baggage that affected the nature and scope of their activities. Religious themes of the chosen people and Manifest Destiny were the ideological fuel that spurred them on in the face of native opposition and an often intimidating natural environment. The majority of settlers imagined themselves as the agents of a crusade with a view to both erecting their civilisation and restoring an earthly paradise.

Some features of the frontier, however, were not quite so malleable. Transportation, communication, and governmental oversight linking the West to the Eastern settlements were costly and unreliable, improving only very slowly over many decades. These circumstances increased settlers' self-reliance and partly shielded them from Eastern interference. Continuing new arrivals of new waves of settlers may have disrupted this shared experience, not least because many pioneers had diverse ethnic backgrounds. But the ideology of republicanism was ready to take over when religious or other cultural traditions were beginning to fragment. '[T]he experience of the developing frontier West, where challenges of distance, isolation, hardship and misfortune helped mould a common response that ignored national origin' (Burchell and Gray 1989: 141) provided the foundations of an emergent cultural identity. The 'loosening social bonds' (ibid.: 135) not only gave rise to proto-democratic practices and self-organising neighbourhood associations but also laid the groundwork for the emergence of the libertarian individual insisting on freedom from governmental interference. Conquering the West was, in Braudel's words, 'a great adventure, both material and psychological' (Braudel 1995: 470).

Capitalism

Besides religion and frontier conditions, some scholars have emphasised the centrality of capitalism in the making of the New World, arguing that capitalist commodification of nature has been the driving force of environmental change (Worster 1994b; Beinart and Coates 1995). "Capitalism organised this great move Westward", proclaims Braudel (1995: 470) and describes the life of an average settler: 'the settler was not a peasant or farmer, rooted in the soil: he was a speculator. He brought off a coup [...]. He gambled; and of course he did not always win. But he went on.' Cutting down the forest and bringing the land under the plough did not require any specialised skills. For a while, settlers could live on canned food, await two or three profitable harvests, and then sell the property, only to repeat the same process somewhere further West.

The quest for economic profit continued to strengthen as the West underwent its conversion to the 'bonanza frontier' in the 1850s. To make one's fortune on the agricultural frontier, settlers were still required to invest several years of hard work in lonely and often dangerous places. By contrast,

> [t]he bonanza frontier offers the prospect of immediate and impressive economic benefit for a relatively low capital outlay; in effect, such a frontier condenses into a brief term the expectation of profit that the agrarian frontiers [...] would require a generation or more to achieve.
>
> (Slotkin 1992: 18)

Given that economic motives for joining the frontier were compelling, it becomes easier to understand the convergence of religious abhorrence for wild nature, the Puritan ethic of hard work, and the quest for rapid prosperity. Adding to this the Smithian conviction about the collectively beneficial effect of the pursuit of individual interest, American society reached a point where both Protestantism and republicanism militated in favour of intensive exploitation of natural resources. Finally, complementing this twofold justification with the traditional European preference for the pastoral landscape produced a powerful rationale for the commercial subjugation of nature. Any alternative stewardship ethic seemed hopelessly outgunned.

The fur trade, which had effectively exhausted its resource base by 1830 (Arnold 1996), and large-scale deforestation are commonly cited as illustrations of the exploitation of nature. In the face of disappearing wildlife and indiscriminate clear-cutting, a first semblance of conservation consciousness emerged. But it would be mistaken to see this as part of a Romantic counter-movement which I describe later on. Rather, efforts at wildlife conservation mostly came from the middle and upper classes whose sports hunters were concerned about the shrinking number of animals (Beinart and Coates 1995). This position, however, clashed with the customary right of settler hunting, and the idea of placing restrictions on the economic activity of free citizens went against the grain. As Worster (1994b: 73) comments,

> [t]o criticize market hunting was to criticize [...] 'capitalist democracy', the notion that the free, egalitarian, and unregulated pursuit of wealth must lead to the welfare of all. A world devoid of wildlife, however, was what capitalism really produced. Capitalist democracy was biocidal.

The US thus harboured an anthropocentric culture which frequently neglected its own dependence on a (seemingly plentiful) natural environment. This demonstrates a fundamental separation between a human realm that encompasses many topics of potential moral concern – such as issues related to health, happiness, and the moral status of individuals – and the realm of nature. Despite a growing worship of wilderness which I discuss later, the separation of the two spheres has historically become more pronounced, economic demands more boundless, and technology more powerful.

An environmental history of the 'Old World'

The European continent has gone through similar phases of environmental and cultural change, but there are many important differences as well. These are both ideational and material: they concern beliefs, values, and customs of European populations as well as the physical transformation of the continent's landscapes over the centuries. Although it is not as developed as its American counterpart, the literature on European environmental history generally reflects this disparity. Most authors in this field adopt a restricted focus on national histories, but this does not mean that intra-European diversity is too overwhelming. While there is a north–south divide in European environmental history, it is tempered by an underlying structure of interaction, borrowing, socio-economic similarities, and visible signs of 'intellectual cohesion' (Roberts 1998: 96). Despite a slower pace of change in Europe, dominant meanings of nature have equally evolved over time. The material basis of human existence necessarily had an influence on ideas of and feelings towards nature. This is most evident in the Middle Ages, where the conceptual work of imagining was rooted in the pragmatic 'use-value' of the natural world and expressed through supernatural 'deification'. In the following, I draw on this dialectical understanding by considering both geographical narratives of physical change and the influence of new ideas and belief-systems on human practices.

From today's vantage point, it is understandable that previous millennia of European civilisation have become idealised to some degree and are often regarded as less anthropocentric and more representative of environmental sustainability. The central issue, however, is not whether or not humans have moulded their environment but on which terms they have interacted with the natural world and which motives lay behind their actions. In the ancient world, it was generally assumed that humans were part of nature. The polytheistic cosmology of Greeks and Romans readily connected their deities with natural phenomena. There was little sense of the vulnerability of the natural environment because the idea of the universe was informed by the concept of a 'grand design', 'with all parts well in place and adapted to one another in an all-embracing harmony [...implying] stability and permanence' (Glacken 1967: 148). Similar conceptions had long characterised the spiritual worlds of 'barbarian' and vernacular cultures. This began to change during the Middle Ages and the two major causes were technology and Christianity (Glacken 1967; White Jr. 1972).

Shifting patterns of land use and population

The modus operandi of medieval communities was centred on the 'use-value' of nature and their environmental identities were largely based on 'practical experience' rooted in 'common sense' and an 'instinctive environmental consciousness' (Myllyntaus 2003). The most widely used record of human-induced environmental change has usually been the pace and extent of deforestation. The Mediterranean region was subjected to millennia of piece-meal clearings of woodlands. There was to be no respite, except for the northern margins of the region, as population levels soon recovered from their previous lows.[4] By the twelfth century, forests were being reduced to 'isolated fragments' (Hughes 2005: 71). In central and northern Europe, clearings on this scale were genuinely unprecedented. From the sixth century onwards, forested areas gradually shrunk, although the most intense onslaught had to wait until the eleventh to thirteenth century (Deléage and Hémery 1990). In France, forest clearings began to change the face of the landscape to the extent that many of the once great forests lost their ancient names and merely became a wooded patch among others (Bloch 1966).

Even more dramatic changes were underway in central and eastern Europe, where regions once sparsely settled now accommodated substantial numbers of migrants from Western areas. As settlements progressively moved into the higher regions of central Poland, rising demand for timber and firewood spelled disaster for low-lying settlements on the river banks, which were threatened by unpredictable floods stemming from the deforestation of mountain ranges (Dunin-Wasowicz 1990). Such settlement patterns have prompted some historians to draw an analogy with the American frontier. After all, the movement of people implied a rapid and profound environmental transformation (deforestation, marsh draining, river control), and, socially, the settlers were leaving the static structures of the feudal West for the non-feudal East where their freedoms were much greater. But significant differences undermine this analogy. As Glacken (1967: 290) argues,

> [eastward expansion], with a modest technology, was the work of centuries, the other [Western frontier], with the aid of machinery, the work of decades. A more profitable comparative history would be with the Mediterranean or the Chinese cultural landscapes through forest clearance, the driving back of wild animals and the extinction of the larger predators, town building, irrigation, canal building, diking, and the like.

European history thus resembles the development of other contemporaneous civilisations.[5] Settlers moved into new areas to stay permanently and they bore the environmental burden of excessive degradation on the mountain slopes. This was quite unlike the fast-moving frontier of the American West, driven by missionary zeal, adventurous instincts, and good business sense.

In utilitarian terms, villagers and city-dwellers relied on rivers and streams, pastures, arable land, and forests and coppices. Forests, in particular, were very important to farming societies which needed not only timber for their houses but also vegetation for foraging animals and coppice for firewood. It thus mattered to local communities whether and how their environments were changed. There was no simple technological determinism or enterprising spirit at work that would inevitably usher in a process of escalating exploitation. The unleashing of capitalist–industrialist fervour had to wait until the enclosures and the enforcement of individual property rights. Medieval societies were, on the whole, less effective at the utilisation of nature than the ancient Romans (Thomas 1984), and their spatial expansion in Northern Europe up until the second millennium A.D. was primarily due to the adoption of a Mediterranean-style cereal-based diet which required more arable land and animal manure (Whited et al. 2005).

Ultimately, however, two main elements conspired against a continuation of the growth trajectory: resource scarcity and the arrival of pandemics. The first aspect was sometimes more relative than absolute, insofar as the feudal structure enforced a royal prerogative over many forests (preserved for hunting) or consolidated the power of landlords over their resources (Bloch 1966). Second, the 'Black Death' dealt a massive blow to European societies and led to a sustained decline in population. Especially in southern Europe, it was this combination of scarcity and disease which put a stop to further environmental exploitation (Hughes 2005).

The spiritual dimension

A second dimension of medieval nature–humanity relations is the spiritual meaning given to environmental interventions. The worldview of medieval societies was intensely magical and replete with supernatural creatures of all kinds, inhabiting the 'wildernesses' on the edge of fields, the rivers, oceans, mountains, and deep forests. One could attempt to bargain with these deities or pray for their benevolence, but their oneness with the forces of nature suggested their ultimate power over human ambitions. In early medieval Holland, '[t]he inhabitants, whose deities were [...] weather gods, simply waited until the overwhelming demonic forces of nature retreated of their own accord' (Zwart 2003: 109–110). The transition towards the end of the first millennium A.D. was not primarily based on an increase in human control through technology. Pagan communities were generally characterised by a degree of passivity and fatalism when it came to large-scale environmental intervention. By contrast, Christian practitioners and scholars spearheaded a vigorous campaign to root out old superstitions, end the deification of the natural world, and establish the concept of the hierarchical 'great chain of being', whereby God was at the apex of the universe and humans were half flesh and half spirit, thus presiding over all other beings, both living and inert. A focus on the physical environment itself, including its supposedly immanent deities, was wrong-headed: 'Salvation to eternal life, not happiness within this natural world, is man's final end; nature is a mere background, a transitory setting for an essentially spiritual quest' (Charlton 1984: 3).

The Christian conquest of hearts and minds was ultimately successful, but at a deeper level, the animistic and 'superstitious' cosmology of medieval polytheism never entirely disappeared, not even in today's late modern age. European elites were Christianised quite rapidly, but the common people changed their habits and beliefs very gradually over the centuries. Accordingly, the most active agents of environmental transformation until the late Middle Ages were Christian monasteries. Spurred on by their accumulated knowledge and missionary zeal, monks were instrumental in many forest clearings, draining projects, and the building of roads and aqueducts (White Jr. 1972). However, the monks did not necessarily act in accordance with the wishes of neighbouring communities. Where conflicts arose, rural dwellers would couch their objections in the powerful language of ancient rights and customs (Glacken 1967: 323). By the late Middle Ages, enthusiasm for technology and the rise of Protestantism heralded a new, even more anthropocentric age. As Lynn White Jr. (1972: 171) phrased it: 'The late Middle Ages considered its advancing technology profoundly virtuous, a manifestation of obedience to God's command that mankind should rule the Earth.' A 'more relaxed attitude' (Charlton 1984: 5) to nature in the medieval era gradually gave way to a new age of technological supremacy, albeit never without prompting significant countermovements. One defining concept of the Middle Ages, a 'balance of power' between the natural and human spheres, would irrevocably recede: 'until the massive industrialization of Europe [...] the relationship to the ecosphere was conceived as an "exchange" with natural forces, often regarded as sacred in a series of myths and religious cosmologies, rather than as a "transformation of Nature"' (Deléage and Hémery 1990: 27). The classical and medieval experiences of nature were ultimately similar: humans were part of 'cosmic nature' and, by extension, whatever they did to the natural world would affect them as well.

Early modern Europe: Realities and reactions

At the beginning of the early modern period, commonly defined as the epoch between 1500 and 1800,[6] most of the assumptions of medieval cosmology were still in place. Christian dogma may have condemned the old animist practices and 'superstitions', but priests were often forced to tolerate them alongside official rites. Magical secrets and concealed patterns of fate were central to the life of medieval country folk. Even the Holy Roman Emperor, Rudolf II (1576–1611), had a keen interest in the occult sciences, including alchemy, and offered to be the sponsor of travelling magi and astrologers. For the vast majority of people, plants and animals were an integral part of an enchanted universe of symbols and charms. 'Crucial to these practices', writes Thomas (1984: 75), 'was the ancient assumption that man and nature were locked into one interacting world. There were analogies [...] between the species, and human fortunes could be sympathetically expressed, influenced and even foretold by plants, birds and animals.'

Scarcity and sustainability

This richly spiritual world retained a powerful hold on the popular imagination. What it did not achieve, however, was the restoration of sustainable relations between humanity and the environment. The principle of utility predominated across all social classes and did not usually collide with any spiritual injunctions, unless a particular species, like the robin or the swallow in England, had been singled out for special protection (Thomas 1984: 76). What limited environmental exploitation in this period were not so much the power of ideas or norms as fluctuations in population. The central Malthusian assumption of much environmental history is an explanation of resource trends through population figures. Grain prices, for example, were always intimately related to the scale of new land clearings and to population numbers (Whited et al. 2005). By the early sixteenth century, when populations had somewhat recovered, many European areas still seemed relatively 'untamed', with 'stretches of wasteland or forest surrounding cultivated areas' (Clout 1998: 225). Deadly diseases, such as plague and malaria, were widespread and malnutrition and local famines were not uncommon. The expansion of cultivation in the countryside, especially onto marginal lands, brought about a steep rise in soil erosion (Hughes 2005). Nonetheless, such 'productivist' measures enabled a considerable increase in population until the mid-seventeenth century. The introduction of novel crops from the New World turned out to be very significant. For instance, maize and potatoes could supply up to three times more food calories per acre than grains.

Deforestation seemed to require a particularly urgent regulatory response. When prices for firewood began to exceed the price of food, one option was to enforce strict regulations, as villages throughout Europe had practised for centuries. John Evelyn's book *Sylva* (1664) and the French Forest Ordinance of 1669 represented a watershed in this regard. For Glacken (1967: 485), they marked 'the beginning of a more reserved attitude toward the modification of nature', and they may have been the product of a long pedigree of conservationist 'propagandists' (Thomas 1984: 198). These regulatory developments were mirrored by similar decisions in other parts of Europe. Many German states were pioneers in this area: between 1482 and 1700, around 151 forest regulations were drawn up (Glacken 1967). The most important insight is not so much the introduction or partial effectiveness of such regulations, but the absence of similar measures in great sea-faring powers such as Spain, Britain, and the Netherlands. The ability to source timber from their colonies allowed these countries to avoid the painful adjustments of rationing and statutory forest protection (Radkau 2005). The legacy of forest history, of scarcity, and self-sufficiency constitutes the economic and ideational context out of which the concept of sustainability would eventually emerge.

The moralisation and 'scientification' of nature

Besides the material and economic driving forces of this historical evolution, powerful ideas actively supported these trends. Christian ethos and religious language were central because they occupied a similar position to science in the modern world – the most authoritative means of justification and consensus-building. A genuine 'knowledge revolution' did not occur before the sixteenth century, but when it happened, the ramifications were far-reaching. The Protestant Reformation weakened the authority of old religious norms. Truth could now be found by reading the Bible and, as some believed, this practice could be extended to the 'book of nature'. Paganism was also dealt a final blow and became subject to intense persecution. One of the Protestant front-runner countries, the Netherlands, vividly demonstrated how crucial the new ideational context was. For instance, a series of devastating floods in the fifteenth century were framed as the deserved 'downfall' of a corrupted medieval world and as the rise of resplendent Protestant era.

> A new type of human being emerged, the protestant self-made man, dwelling in a self-made environment. The natural world, the inhospitable semi-maritime realms of old, were now radically disenchanted and *Kreaturvergötterung* (deification of nature) was vehemently rejected.
>
> (Zwart 2003: 114)

Protestant disenchantment was never complete and one should not ignore the experiences of more conservative, staunchly Catholic nations in the Mediterranean region. But Protestant ideas were buoyed by the new sciences and their optimistic practitioners.

Given this book's focus on attitudes towards nature, the emergence of a new class of natural philosophers or scientists is of genuine significance. Without them, the realms of nature and humanity could not have been conceptualised in now customary ways, and without their inventions human control of the environment would not have reached the level of today. The seismic shift took place between the fifteenth and seventeenth century, and it ushered in a new 'Western' trajectory of thought, 'marking itself off from the other great traditions' hailing from world regions such as India and China (Glacken 1967: 494). The traditional conception of the 'great chain of being' was becoming increasingly untenable. Copernicus and Galileo were the first to question the centrality of humankind in the cosmic order by advancing a helio-centric theory of astronomy. Descartes and Bacon went further still by insisting on a disembodied form of rational thinking and the introduction of a thoroughly mechanistic worldview (Whited et al. 2005). These natural philosophers complemented the Protestant disenchantment of the natural sphere.

> In place of a natural world redolent with human analogy and symbolic meaning, and sensitive to man's behaviour, they constructed a detached natural scene to be viewed and studied by the observer from the outside, as if by peering through a window, in the insecure knowledge that the objects of contemplation inhabited a separate realm, offering no omens or signs, without human meaning or significance.
>
> (Thomas 1984: 89)

These 'scientific naturalists' fashioned a new sense of 'nature' which had never before existed in human history. It is captured by Williams' (1976) third 'material' meaning of nature (Chapter 3). The new notion of nature was not only distinct from the human sphere but also systematic. It could be studied and understood through human faculties and was malleable by human hands (Charlton 1984).

The momentous changes underway could not have succeeded to the same degree if they had not found allies within the religious and secular establishments, especially at a time when the moral order underpinned by the older animated realm of nature and its associated customary rights was increasingly destabilised by 'a political economy which transmuted such customs into cash-values and denied the godly origin of the soil's fertility' (Schaffer 1997: 125). In the seventeenth and eighteenth centuries, when Britain's elites were beginning to carve up the common lands and create enclosures, traditional land-use practices were defended by a combination of Christian morality and socio-economic imperatives. Bitter contests would be fought over

> a moral economy which celebrated use rights and custom in the name of a profoundly theological account of the sacred bounty of the soil. Food riots were not 'rebellions of the belly', but appealed against profiteers to a principled understanding of the proper means of price information and the transfer of the earth's produce to the market [...]. Crop failures would be seen as acts of God and attempts to profit from high prices as blasphemy.
>
> (ibid.: 125)

Nature was both a spiritual realm and a moral resource for the struggles of everyday life. By appealing to a God-given order, the defenders of traditional rights sought to deprive economic reformers of their legitimacy and brand their plans as fundamentally 'immoral' and 'unnatural'. The political and economic struggles during this period of accelerating change were, in other words, not merely fought on the ground (over forests, coppices, grazing land, communal rights, and obligations). They were equally being waged in the realm of ideas where the sacredness of nature – inspired by medieval animism and updated by a Christian view of natural order – was losing out to newer conceptualisations that claimed to have a superior understanding of both nature and God's will. The 'knowledge revolution' successfully reinterpreted religion as well as humankind's role in the broader cosmological framework: humans ascended to the status of co-creators, perfecting creation rather than debasing it.

The 'scientific naturalists' sought to delegitimise an older understanding of cosmology and transform the 'moral economy' that governed early modern societies. 'Traditional custom was deliberately redescribed as a source of social disorder, superstition and unreason' (ibid.: 128), while the new 'economy of nature' held sway. Influenced by an early swing towards Protestantism and capitalism, in Britain the project of modernisation was most successful, although it represents an approximate model for other European countries as well. The physiocrats in France, including key writers such as Mirabeau or Quesnay, only became influential in the second half of the eighteenth century, but their economistic and materialist outlook was essentially similar (Herlitz 1997). Overall, the early modern period was indeed a watershed era during which environmental 'instrumentalism' (Macnaghten and Urry 1998) rose to prominence by undermining older currents of vernacular and classical thinking. The concept of human dominion became 'sharper and more explicit', whereas the notion of humans as the 'perfectors' of creation relegated religious injunctions to a background role (Glacken 1967: 494–495). These ideational shifts reflected parallel developments at the material level: a 'historical threshold was crossed in the degradation of ecosystems' as European societies moved through various phases of agricultural 'marketisation' towards a full-fledged version of industrial capitalism (Deléage and Hémery 1990: 29).

The modern era

The entanglement of material and ideational factors is characteristic of a dialectical historical process. If new ideas enabled the objectification and more effective exploitation of nature (and human communities), associated material benefits, and the breath-taking power on display have, in turn, lent added intellectual strength to processes of modernisation.

The conceptions of nature and ideals of human progress described above fed into a new consensus among significant sections of Europe's intellectual classes.

Mainstream intellectual views by 1800 and their critics

Until the early nineteenth century, the overall aesthetic and cultural outlook conceived of the natural environment as a well-ordered and pleasant terrain free of uncontrolled wildness. Most writers encouraged efforts to

> win new land from the moors, the fen, the old forests. They gloried in the ideal of a beautiful village resting in well-tilled fields. They had faith in technology and in the possibility of improving the individual advances of knowledge; they saw that nature also could be improved with this new knowledge, itself a product of an awakened curiosity.
>
> (Glacken 1967: 665)

These essentially pastoral images of nature were not dissimilar from the classical age, except for the desire to impose a predictable, human-controlled order. In one of the first articulations of 'human improvement' of plants – and thus supportive of today's biotechnology enthusiasts – one of the writers of this period, Count Buffon, argued that 'bread grains were not a gift of nature; they were a product of experiment and the application of intelligence to agriculture' (ibid.: 676).

Opposition to an uncritical embrace of scientific progress mainly emerged from two quarters. First, from those who took the 'natural order' established by God more seriously and argued that excessive interference was manifestly not part of God's original design[7] (Thomas 1984: 278). Second, a more original critique arose around 1800. It manifested itself in a German current of thought labelled *Naturforschung* and defined itself in opposition to the better-known Romantic *Naturphilosophie*. The former thinkers, also known as Romantic naturalists, did not deny the need for human intervention. They did, however, take on the philosophical underpinning of the now dominant version of the humanity–nature relationship. Their thinking harked back to the late Middle Ages and also corresponded to a growing ecological understanding of interdependency, regress, and collapse. On the whole, however, dissenters were overcome by the rising tide of interventionist ideas, powerful economic interests, and material developments on the ground.

The agricultural revolution

Driven by new knowledge and confidence in the benefits of transforming nature, the agricultural revolution across Europe sought to replicate the advances of industrialisation and provide enough food for fast-growing populations. Both moral justifications and expedience were firmly in place to defend the unprecedented socio-economic and physical changes that define modernity as a historical phase (Kemp 2004). Conditions across Europe differed markedly and the scale and speed of modernisation were far from uniform, but the direction towards the most successful model practised in the Flemish region – with animal fodder, cash crops, cereals, craft industries, and concentration in towns – was widely taken for granted (Clout 1998: 227). More integrated national economies lowered the risk of local famine and spurred both experimentation and mechanisation, releasing agricultural workers into the industrialising cities. The German states, the Netherlands, and Britain eagerly adopted the new techniques. German agriculturalists, who were predominantly large landholders with a taste for innovation, widely regarded the land as the root of all national wealth. Great efforts were made to increase agricultural output and reclaim marginal lands (ibid.: 231–232). An even more concentrated form of landholding pertained in Spain where poor soils and dry summers acted as barriers to wide-scale mechanisation (Simpson 1995). Similar conditions could be found in other Mediterranean countries such as Portugal or Greece. However, the agricultural revolution did not pass these states by. It simply materialised gradually – often several decades or even a whole century later. Increases in yield and overall production in the Mediterranean were initially achieved through pre-industrial techniques of expanding cultivation and hiring more labourers (Whyte 1998).

Besides north–south differences, the case of France demonstrates that specific national trajectories shaped the way agricultural innovations were implemented on the ground. Here, ancient rights were thoroughly entrenched and protected; the modernising process was particularly slow and difficult:

> small, owner-occupied properties, local resources, and common access structured agriculture in France. A host of traditional collective rights – to glean in harvested and unenclosed fields, to pasture in common meadows, and to cut wood in forests – had been guaranteed by the Rural Code of 1791 and remained fiercely defended in the nineteenth century.
>
> (Whited et al. 2005: 105)

The defence of traditional ways of life is clearly linked to broader socio-economic questions and the cultural status of peasants as the nation's revered producers and caretakers of the countryside. Especially this latter cultural role came to the fore in the twentieth century and takes a central place in the national debates over GMOs (see Chapters 4 and 6).

Ideals of technological mastery over nature and economic interests were the heart of the agricultural revolution, but looking at the statistics for the eighteenth century, one is struck by an ecological precipice opening up due to population growth, deforestation, and soil erosion (Whited et al. 2005: 91). Figures for forest cover in 1868 show a paltry 12% for Sardinia, 9% for southern Italy, and 4% for Portugal. Secondary effects such as flooding, soil erosion, and siltation of rivers were rampant (Hughes 2005: 118–119). In northern Europe, environmental impacts were no less devastating. Hungary had lost close to 75% of its forests by the early twentieth century (Kiss 1990). In Denmark, forest cover decreased from 20–25% in 1600 to 8–10% by the mid-eighteenth century, and the speed of deforestation in Germany exceeded a sustainable rate of wood consumption by 50 times before more effective rules of forest management were imposed (Whited et al. 2005).

Three developments 'rescued' European societies from ecological collapse and Malthusian crisis. First, afforestation was stepped up dramatically and heavily relied on fast-growing monocultures which fitted well with the dominant mechanistic worldview. Second, for some European societies, the 'ghost acreage' of colonial possessions and the high seas offered opportunities for fishing and whaling on an industrial scale as well as providing foodstuffs, cotton, and furs in great quantities (Whited et al. 2005: 93–94). The third short-term 'safety valve' were successive waves of emigration and a huge expansion of cultivated areas.

The anchor of 'nature' in a utilitarian age

Historically speaking, transformations and revolutions tend to breed reactions and counter-revolutions. Europe's agricultural revolution and associated upheavals entailed a far-reaching commodification and rationalisation of nature which clashed with millennia of humanity–nature relationships marked by interdependence. The reaction manifested itself in a new wave of appreciation for nature. It encompassed an affective and aesthetic dimension which was, at times, wholly divorced from use-value and had never before gained comparable prominence. Yet, the precursors to nineteenth-century Romanticism are all too often forgotten, even though the pastoral aesthetic of the Renaissance continued to coexist alongside modern innovations (Walter 1990), and, in Europe, it has remained the mainstay of popular feelings for nature to this day.

A promising mechanism for understanding the ongoing debate over nature and humanity is the notion of identity. In particular, 'environmental identities' are rooted in both material–utilitarian and affective–cognitive interactions with the land, yielding a cultural pattern of practices and attitudes. Changes in landscape or environment do not solely affect small circles of ecologically or aesthetically minded intellectuals. They are often seen as disturbing people's everyday lives and moral understanding of the world. In the following sections, I trace the historical development of aesthetic and identity-based counter-currents and suggest that, in Europe, the 'naturalisation' of and national identification with productive landscapes has made the accommodation of industrial modernity a more conflictual affair. To begin with, however, I return to America's cultural and environmental history and trace the emergence of a proud nation enamoured with potent symbols of nature.

America: Exceptionalism in the making

The nineteenth and twentieth centuries echoed with the voices of those who protested against machine-led industrialisation. In confrontation with the Enlightenment, the Romantics praised 'irrational' qualities like intuition and creativity and popularised a new awe-inspiring conception of beauty – the sublime. The idea of wilderness itself was not fundamentally altered, but it was now becoming a desirable setting: 'wilderness was [not] any less solitary, mysterious, and chaotic, but [...] in the new intellectual context these qualities were coveted' (Nash 1967: 44).

The origins of Romanticism

The Romantics had their original strongholds in the Old World (particularly Germany and Britain), but as far as raw nature was concerned, there was no better canvas than the American continent. Romantic disciples were found predominantly among East coast literati and artists. Those who worked on the land – including the vast majority of those inhabiting the frontier regions – appeared to be immune to such appreciation of nature. Their mindset was still focused on the subjugation and exploitation of a wilderness perceived to be hostile to human civilisation. Living off the land made it impossible to exclude pragmatic and economic considerations.

Elements from history, religion, and modernity help to differentiate American Romanticism from the European strand. First, many commentators have remarked that American culture was strangely devoid of historical context. Its 'manifest destiny' was a future-oriented ideal and influential figures like William Gilpin revelled in the 'pastlessness of the New World' (Stoll 1997: 119). This sense of relief was shared by many Americans, but subconsciously a great affective void began to take hold, expressing itself in feelings of inferiority towards the rich history of the Old World. Ruined castles, monasteries, and other traces of ancient kingdoms were greatly attractive to the Romantic imagination, but these were entirely absent from the American landscape. In addition, the lack of a shared sense of history presented a serious problem for national unity (Shepard 2002). Nature and its many wondrous manifestations offered themselves as a cultural fulcrum. As explorers and settlers rode through the West, they passed by extraordinary stone formations and began to envision figures from the European or the Biblical past, the remnants of ancient civilisations or monuments erected for the Almighty himself (ibid.: 241–242).

Alternatively, the sublime beauty of wilderness could also be employed to argue for the superiority of the American continent. The dismissal of ancient art and urban refinement was strengthened by a form of religious dualism which became the guiding logic of Romantic thought. Wild nature, true wilderness, was where spiritual fulfilment and sustenance could be found, far away from the depravity of the city, its overly refined civilisation and foul-smelling factories. The privilege of embodying the virtuous opposite of the city was also being claimed by the physiocratic tradition and the defenders of rural society (Tuan 1974). The turn away from the 'fallen' city, which persists to this day, thus came in two versions. The wilderness path was preferred by the Romantic elites, while the rural, and later preferably suburban, solution has become the common sense of large parts of the population. Conversely, in Europe, in large part due to the absence of wilderness, the rural ideal has prevailed, albeit with less vehemence.

Transcendentalist dualism

America's fervent religious heritage distinguishes it from the broadly secularising character of European modernity. The intellectual movement that most influenced the Romantic turn in the mid-nineteenth century was American transcendentalism, and especially its major spokespersons Emerson and Thoreau. The latter, however, was an ambivalent thinker who could equally be seen as a 'counter-cultural' source for the alternative project of bridging the divide between wilderness and civilisation (Nash 1967). In his best-known work *Walden* (published in 1854),

Thoreau 'mythologizes the farmer to justify his practice and mode of agriculture, and to reconcile the tensions between wilderness and civilization through a "middle landscape" of farming' (Scheese 2002: 46). These views are largely analogous with European conceptions, as discussed below. Emerson, on the other hand, encapsulates the problems of idealism which have afflicted the Romantic defence of nature to this day. He tried to rescue nature from the image of the 'howling wilderness' by locating a presence of divinity within it (Nash 1967). Yet, Emerson hardly felt any concern for the well-being of wild nature. Coates (1998) thus portrays Emerson as a firmly anthropocentric thinker who was, in contrast to Thoreau, an early admirer of Francis Bacon and a staunch believer in 'civilisation' and progress. The separation of humankind and nature was never questioned: 'Through solitude in nature', Emerson ruminated, 'we recognize our apartness and our essence' (ibid.: 135). If beauty and a sense of divinity could be found in nature through a mystical relationship with it (Stoll 1997: 107), it was nevertheless not wild nature that was celebrated, but its capacity to produce a sense of *human* divinity. In other words, nature was 'a conduit, a raw material to assist the human spirit in its quest for perfection' (Coates 1998: 136).

Transcendentalists still retained an appreciation for the pastoral landscape, but the emphasis on purity and spiritual qualities accelerated the drift towards a higher expression of nature – that of wilderness. The aesthetic battle was being waged between the 'sublime' – fusing the feelings of fear, awe, and wonder – and the more classical 'picturesque' which suggested a much kinder, tamer beauty marked by apparent human purpose (ibid.: 132). The great 'prophets' of twentieth-century American environment all drew on these venerable traditions of nature philosophy. John Muir (1838–1914), co-founder of the Sierra Club, was one of the most eloquent champions of American wilderness, but as much as he insisted on the sanctity of animals and plants, he always envisaged the moral and spiritual benefits for human beings as the chief aim of wilderness preservation and national parks (Stoll 1997). A comparison with Aldo Leopold (1878–1948) – a pioneering ecologist, but also a keen hunter and farmer – largely mirrors the differences between Emerson and Thoreau. Leopold's major work, *A Sand County Almanac* (1949), had 'many of the same talismanic qualities as *Walden*' (Payne 1996: 123). Leopold began to question the ways in which hunters extinguished unwanted predators and farmers depleted the soil. Rather than lobbying the government to set aside wilderness reserves, as Muir did, Leopold focused on restoring the 'wildness' of productive territories, calling for 'a land ethic in which the private landowner would recognize that in the long run it is better – aesthetically, morally, and economically – to act as steward of one's property' (Scheese 2002: 98). Nonetheless, Muir's friendship with President Theodore Roosevelt, the relative success of his campaigns, and Leopold's despair at the destructive farming practices of (virtually all) his neighbours indicate the prevalence of a 'bifurcationist' civilisational attitude in the US.

Consequently, the fledgling US conservation movement did not harbour the anti-industrial convictions of many other twentieth-century environmental movements. On the contrary, it saw science as a medium of human ingenuity (Beinart and Coates 1995) and, in the US, science has retained much of the positive aura which, across the Atlantic, was overshadowed by European metaphors of Frankenstein. Luddism and the defence of time-honoured customs, landscapes, and ways of life has been a characteristically European response to the industrial age and the rationalising spirit. Americans were never quite as close to becoming disenchanted with modernity (Coates 1998: 136). The material transformation wrought by modernity did not worry them nearly as much as its de-spiritualising consequences. These feelings joined a long-standing sense of an exceptionalist destiny and a growing desire to break away from the perceived cultural superiority of the Old World. Nationalism was the major social force deployed to fill the affective void and boost a genuine American identity.

The search for identity and distinctiveness

There are good grounds to argue that the socio-cultural and emotional consequences of modernity propelled the American people towards a reconceptualisation of wild nature. The lack of a 'deep' history and the increasing diversity of American settlers were both a blessing and a burden. Americans were subconsciously yearning for a unifying myth that went beyond the lofty values of liberty and the pursuit of happiness. America's destiny was to be a free and prosperous beacon for the world, but it was hard to maintain the faith without the solace of a 'golden age'. The aesthetic symbolism connecting the past with the present would be the sacred places of wilderness, standing in defiant contrast to the architectural wonders of the Old World. The first step towards a distinctive national identity was the mythical reconstruction of America's short history since the first settlements and the idea of the Western frontier played a major role in the making of 'Americanness' (Burchell and Gray 1989: 140).

The two major mythologists of the frontier were the historians Frederick Jackson Turner and Theodore Roosevelt (US President

1901–1908). Their different interpretations of the historical significance of the frontier for America's future both expressed a widespread anxiety about identity at a time of rapid change (Slotkin 1992: 32). Turner's privileged actors in the historical evolution of 'Americanness' were the settlers living and working on the agricultural frontier. Roosevelt felt greater appreciation for the trappers and hunters, since these were men who represented the values of virile virtue and rugged individualism (Nash 1993). Roosevelt's version of frontier history turned out to be a more influential force in the American imagination. His hero-centred narrative provided ready-made content for the developing mass culture (Slotkin 1992: 61). Important as these two intellectual figures were, they are also an expression of the historical evolution of cultural and political attitudes. Heralding a fundamental cultural shift, it gradually transpired that the currents of nationalism and Romanticism were joining together in a great movement towards a distinctive American identity. Wild nature was an attribute that could rekindle spiritual feelings and existed in abundance in the New World. Nothing else could set America apart from the Old World in quite the same way (Nash 1993).

An incidental aesthetic transformation

The strength of Romantic nationalism is linked to a concomitant shift in aesthetic perceptions. The nineteenth century witnessed a transition from older pastoral and picturesque ideals of beauty to newer ideals of the sublime and the wild. The first generations of American settlers were still thoroughly European in terms of cultural outlook and the rigidly religious world of Puritan farmers led many to confer a deeply negative meaning on 'wilderness'. The later shift from the pastoral garden to wilder parkscapes was an aesthetic revolution; although, it drew its inspiration from English parks which set the island's aesthetics apart from continental European gardens. The transition of American preferences towards wild nature preserves was not completed before the early twentieth century. In fact, the first American national parks, Yosemite (1864, formal status in 1890) and Yellowstone (1872) were, for a significant period, still visualised as paradisiacal garden landscapes. The pastoral ideal had been running strong in much of the world for at least several millennia. The garden and the rural, pastoral landscape were the preferred settings throughout the classical period up until today, with genuine innovations only occurring since the European Renaissance. Italy pioneered the geometrical, formal garden often adjoining villas. Over time, the Italian ideal spread throughout Europe and culminated in French garden design. At Versailles, the stylised, geometrical garden represented the 'triumph of culture over nature' and was 'informed by a mechanistic conception of nature as a well-regulated and predictable system' (Coates 1998: 117).

It was the Romantic movement which shattered this rationalist-aesthetic amalgam, and Germany and England were among the countries where its impact was most strongly felt. English gardeners and aestheticians were in the 'vanguard of reaction' (ibid.) to the geometrical model. Newer English gardens allowed for a scenic view, imitating a romanticised pastoral landscape that was often dotted with manufactured classical statues. But what seemed to many French observers as a 'wild' garden, a *jardin anglais*, looks rather tame and arranged to the contemporary eye. In America, the banner of the English aesthetic was carried by Thomas Jefferson and his followers. Jeffersonian philosophy became associated with the pastoral ideal of the yeoman farmer, harbouring feelings for a cultivated nature that were quite similar to the European taste (ibid.: 123).

The convergence of nation and wilderness

Thus, the tame pastoral landscape initially remained the dominant aesthetic vision in the New World. Americans adopted the logical extension of the garden, the much larger park landscapes, to raise their own cultural profile. The first two national parks, Yosemite and Yellowstone, reflected a combination of English aesthetic judgement and American democratic republicanism (Beinart and Coates 1995: 74). They did not represent an attempt to revolutionise the idea of nature, but they revealed the ambition of improving what the English had achieved. The sense of improvement arose from the republican values underpinning the American parks. Frederick Law Olmsted, the architect of the New York Central Park, was an influential figure among those pressing for Yosemite to be designated as an inalienable public parkscape (Olwig 1996). Olmsted's republican case undoubtedly helped to protect Yosemite from exploitation, but this could not have happened without a general aesthetic mood in favour of landscape parks. Yosemite and Yellowstone were touted as refined parkscapes on a much grander scale than in the Old World. Those who had long felt culturally inferior to the Old World could rejoice because this landscape garden scenery more than rivalled English gardens, while also providing universal, egalitarian access.

Simultaneously, however, fundamental cultural and political changes were underway that would make these aesthetic motives increasingly unattractive. By the middle of the nineteenth century, the differences with Europe had grown to a level where the search for distinctiveness and even exceptionalism was becoming a national passion. Advertising the democratic quality of the American parks may have been a first step, but many people yearned for a grander statement. National parks thus began to symbolise 'a powerful cultural statement fusing notions of nature and nation', a quality they have retained ever since (Beinart and Coates 1995: 90). From the great intellectual pool of cultural nationalism, religious messianism and Romanticism arose a new ideal of nature as a moral and spiritual force. In addition, the defenders of national parks soon realised that most Americans had little love for unproductive rural landscapes that appeared to be waiting for the plough (Olwig 1996). This provides the final reason for why the shift towards wilderness values took place within only two decades after the official protection of Yosemite National Park in 1890. In summary, the forces of conservationism and Romantic nationalism were both pushing towards a more exceptionalist set of wilderness values. The dozens of parks that followed Yosemite and Yellowstone slowly abandoned the imagined pastoral beauty of the two pioneers. Olwig (1996: 398) offers Mount Rushmore (established in 1925) as an iconographic counterpoint. It belongs to a later era 'when barren cliffs did not need to be contrasted with fertile meadows in order to conjure up a picture of nature'.

At the dawn of the twentieth century, American conceptions of nature had thus undergone an aesthetic transformation and acquired an unprecedented level of moral and patriotic vigour. They also provided a congenial cultural context for the objectives of competing cultural and political movements. While Romantics could seek out communion with nature in national parks, industrialists could go about the business of resource exploitation as long as they did not threaten a 'spiritual' sanctuary or an iconic haven of 'wild' nature. The middle landscape of agricultural and pastoral America was being squeezed between the two poles and lost some of its moral superiority over the city to the new wilderness ideal. Neither the middle landscape nor the wilderness reserves, however, could in the long term escape the relentless advance of capitalist economics. Agricultural areas were progressively turned into 'food factories' concerned with maximising output and profits, whereas national parks saw steadily increasing number of visitors which threatened to overwhelm their fragile wilderness experience (Beinart and Coates 1995: 78). Economic commodification and the conceptual division of humanity and nature embedded in religious and Romantic thought coalesced. Their product was a bifurcated American landscape.

Europe: The visceral value of 'ordinary' landscapes

Identifying the root causes of cultural trajectories is a difficult enterprise, but a number of historical shifts have undoubtedly been central. One of them concerns basic American cosmological beliefs and civilisational dispositions derived from Christian-European ancestry. Tuan (1974: 247) has articulated this position by arguing against the distinction between adaptation to and domination of nature:

> A truer distinction is to recognize that [...] traditional peoples lived in a vertical, rotary, and richly symbolical world, whereas modern man's world tends to be broad of surface, low of ceiling, nonrotary, aesthetic, and profane.

Accordingly, as I argued earlier, the late Middle Ages witnessed the gradual decline of spiritualised, enchanted nature. After 1500, a new way of experiencing and conceptualising the world began to take hold. America's frequently Puritan settlers acted as a radical vanguard of this broader movement. The powerful idealism of the Romantic movement did not thrive on blank slate; hence its relative compatibility with existing ideational-religious currents and the civilisational disposition of American society. With Europe as the origin of most American settlers, one might assume that European societies have been subject to similar cultural shifts. An undeniable overlap with certain trends in American history could be taken to mean that the two continents underwent a parallel evolution. Against this reading, most of the following observations emphasise the alterity of the European trajectory.

An emerging taste for nature: Romantic revolt

In some countries, such as Britain and Germany, Romantics went beyond the cautious embrace of the wild that is evident, for example, in Jean-Jacques Rousseau's writings. European Romantics often sought to escape both the stultifying effects of the picturesque countryside and the mechanising furore of industrialisation. It is imperative to stress that, despite Romanticism's long-term influence on environmentalist thought, this ambition only united a minority of upper- and middle-class poets and literati in various European countries. As Charlton (1984: 59) notes,

> [w]hat the minority of eighteenth-century people [...] managed to achieve was so to enlarge their appreciation of nature that they

> could inwardly accept nature's ultimate power over us. [...] [T]hey felt almost strengthened by their sense of that power: hence the near-exultation, the mood of almost transported affinity which they sometimes expressed.

Exultation and sublimity carried distinct religious or mystical connotations, thriving on a pantheist understanding of the world and finding spiritual sustenance in a nature seemingly unsullied by human purposes. 'Nature was not only beautiful; it was morally healing' (Thomas 1984: 260). Mountains, forests and oceans were 're-enchanted' and celebrated for their sublime beauty and spiritual associations. To a significant extent, the new valorisation was made possible by the growing strength of natural theology. This philosophical school originated in seventeenth-century Britain and sought to counteract the pessimistic Christian reading of nature that was dominant in the Middle Ages. Before the eighteenth century, anything outside the sphere of 'usable' nature had been regarded as sterile or dull (Charlton 1984). A new generation of natural theologians held that God's purposeful design was evident in all features of the land and that these might all be regarded as beautiful in their own right.

Besides the preparatory groundwork laid by natural theology, European Romanticism also arose in reaction to the cultural and economic transformations that accompanied the era of modernity. The disenchantment of nature now coalesced with a 'tremendous escalation in the technological manipulation of nature' (Bess 2003: 117). Even more than in America, this process engendered feelings of anxiety. Wild nature, as an aesthetic and spiritual counter-model, offered itself as a spectacular alternative, in terms of mythical power and distinctiveness. On the one hand, '[n]ature in its wilder forms served as a stick with which to beat the idea of progress and the worship of economic growth' (Beinart and Coates 1995: 94). On the other hand, it could function as a means to re-enchant a world increasingly marked by rationalisation and industrialisation. Romantic poets, such as Wordsworth and Coleridge in Britain or von Eichendorff and Goethe in Germany, represented the vanguard of what was both an aesthetic and a broader cultural revolution.

By the late eighteenth century, in Britain, the ideology of progress with its enthusiastic support for clearances, draining, and large-scale agriculture was essentially running out of unconquered territories. The last wave of enclosures had entrenched a patchwork landscape of hedgerows, hawthorns, and the occasional group of trees. For a time, beauty and utility appeared almost reconciled. Yet, in order to distinguish themselves from their neighbours, as early as the seventeenth century, noble families would plant so-called wilderness areas – avenues of trees or patches of forest – on their lands, and deer parks became 'increasingly ornamental as owners demonstrated their wealth by deliberately rearranging the landscape' (Thomas 1984: 202).

In southern Europe, Romantic ideas and sensibilities were markedly less successful. The classical vision of the 'natural' landscape, which animates pastoralism, has continued to dominate, although it also accommodated some images of wild nature. As early as the late Middle Ages, statements of an updated classical vision can be found in the memoirs of Pope Pius II (1405–1464), which

> combine a love of scenery with historical associations, seeing in the fusion of beauties of landscapes altered and unaltered by man. [...] There is wildness in his Italian landscapes, but reminders of human activities – the olive grove, the vineyard, the ruin – are never far away.
> (Glacken 1967: 355–356)

Images of nature in contemporary southern Europe still resonate with these reflections. Catholic societies, in particular, were not predisposed towards 're-enchanting' nature, after having fought medieval animism for centuries. Moreover, the concept of nature was often perceived as a covert attempt at establishing a secular morality in place of a religious order. That was indeed the project of many French *philosophes* of the Enlightenment (Charlton 1984: 6–7). The 'tyranny' over nature epitomised by the geometrical gardens of Versailles was gradually rejected in favour of slightly 'wilder', 'English' gardens, but these were all relatively timid steps. Even Rousseau, commonly counted among Europe's proto-Romantics, would not venture beyond the multi-faceted harmony of a 'gentle, rural world' (ibid.: 39). Southern European countries never adopted the Romantic adoration of wildness to the same degree. Mediterranean societies as a whole retained a thoroughly integrated mode of the humanity–nature relationship.

Reconnecting nature and society: The pastoral revival

Ultimately, Romanticism stole the limelight from what turned out to be a more enduring opposition to the modern industrial trajectory, rooted in national cultural identities and the lifeworld. The pastoral ideal of tamed, humanised, and beautified nature united all those who felt unease at modernity's material and cultural transformations. It rose to prominence in the eighteenth and nineteenth centuries when it became necessary to distinguish itself both from the Romantics and from the champions of agricultural modernisation. To this day, it could be seen as embodying the 'common sense' of most Europeans with regard to nature, in both moral and aesthetic terms. As Thomas (1984: 253) mused, '[t]he ancient pastoral ideal has survived into the modern industrial world. It can be seen in the rural imagery so often employed to advertise consumer goods; and in the vague desire of so many people to end their days in a country cottage.' However, disillusionment with the rising tide of modernity proceeded in slow motion. And paradoxically, the Georgian landscape of 'chequerboard fields, hawthorn hedgerows and tidy coppices' (Porter 2000: 15) – which today is seen as the epitome of nature in Britain – was increasingly idealised, despite being the result of enclosures and agricultural intensification.

Over time, however, the overlap between pastoral sensitivities and utilitarian rationality began to shrink. As ebullient capitalists turned into landed gentlemen, their instincts were tamed and both pre- and anti-modern sentiments attained a central place right within the ruling classes of the time (Wiener 1981). The absorption of the working class into a rural-pastoralist worldview made possible the collective self-image of Britain as a 'green and pleasant land' (Macnaghten and Urry 1998: 176). This ability to combine elements of pre-modernity with images of an alternative future moving at a gentler pace was pastoralism's greatest strength. The ideal of pastoral beauty – drawing on ancient images of fertility, abundance, and stability – could complement the human heritage expressed in nature through a mystical association with the divine. Thus, in Britain, where forests were scarce, 'trees were increasingly cherished, not just for their use, not even just for their beauty, but because of their human meaning, what they symbolized to the community in terms of continuity and association' (Thomas 1984: 214). By the mid-eighteenth century, 'the analogy between groves and ecclesiastical architecture became commonplace' (ibid.: 216). A similar dynamic applied to Germany where a rising appreciation of 'natural' forests reflected their subjection to the requirements of commercial forestry. Radkau (1997) recounts how a new generation of foresters sought to restore the country's woodlands to their former glory. The reformers' predilection for long-term cutting cycles and strong, fully-grown trees stemmed less from knowledge of ecological science than from a desire to restore the kind of woods seen under traditional peasant forestry (*Plenterwirtschaft*) in which only selective cuttings were made. There was a sense that 'a good tree should have a whole life and should die only when it is old. We can recognize in the background a feeling of partnership with nature' (ibid.: 231). Implicit in the new vision of forestry, which was not implemented until the early twentieth century, were cultural assumptions of 'naturalness' and of the environment as embedded in a web of moral relationships. This contrasted with the dominant school of scientific forestry and its focus on efficiency, economic benefits, and monocultures.

These trends signified more than the beginnings of a broader Romantic movement. Nor were they mere reflections of the ancient classical ideals, although they had a good deal in common with them. More broadly, they were part of a cultural, aesthetic, and, occasionally, socio-economic reaction against the juggernaut of modernisation which threatened centuries of traditions laden with cultural and spiritual meaning. Crucially, this was a genuinely widespread and popular feeling, not confined to certain social classes or professional organisations. By contrast, European Romanticism's impact on the general trajectory of humanity–nature relationship was circumscribed by social class and by material circumstances, such as the availability of wilderness.

Nationalism and nature: The unification of elation and belonging

The fusion of nature and nation in American history was not a unique historical process. The surge in national feeling and renewed attention to nature in Europe merged into a deeply patriotic current of nationalist ideology. The significance of this lies not so much in its apparent similarity to American developments, but rather in its ultimate difference – nature, the signifier, is not wilderness – and its role in forging vigorous national identities wedded to 'productive' landscapes. While sharing the broader features of the nationalisation of nature, European countries went their own specific ways according to their varying cultural resources, social structures, and material–environmental endowments.

Emerging forms of cultural nationalism were partly rooted in the fifth sense of nature which I called 'nature as nostalgia' (see Chapter 3). First, the nineteenth century witnessed an attempt to combine elements of Romanticism with new forms of belonging (the nation), thus softening modernity's impact. Second, this process often stoked a craving for the socio-cultural stability of a 'golden past' and a simpler rural life 'in harmony' with nature, which Worster (1994a: 471) calls 'Arcadianism'. Nature becomes a placeholder for biological and social order as well as traditional rural virtue. On the one hand, conceptions of nature and their associated landscapes are capable of grounding relations of belonging or identities in a material form, namely in places of historical or scenic interest, in natural and/or national monuments (Wallwork and Dixon 2004). On the other hand, the 'quest for identity' which had 'traditionally been satisfied by religious belief systems' (Eder 1996: 186–187) cannot do without elements of spiritual transcendence, without the semblance of an overarching natural order. Arcadianism is closely related to what Eder terms the 'conservationist package' in which nature's stable order is legitimised both through divine supervision and scientifically defined biological balance: 'nature not only has a morality in itself which man cannot alter, but it also shares a moral world with the human world' (ibid.: 178). An assemblage of medieval (pagan and Christian) attitudes to nature and modern ecological insights underlies this approach.

To a greater or lesser extent, European variants of cultural nationalism all relied on such collages of national landscape symbols, identity construction, and mythical 're-enchantment'. They often included an 'organic' national past for solace and inspiration, such as the Germanic myth of origin and its connection to the ancient forests. Clearly, cultural nationalism implied a departure from the ancient classical attitudes to nature, even if the still dominant pastoralism continued to revere the humanised, agricultural landscape. In Britain, the dearth of wilderness turned the pastoral landscape into the privileged receptacle of renewed affection. The countryside offered itself as a contrast to urban areas: 'it was a realm filled with landscaped estates, profitable agriculture, concentrated wealth and rural leisure pursuits' (Macnaghten and Urry 1998: 177). However, the modern 'construction' of the countryside ideal has at times been overstated. Britain had long been an essentially rural society founded on governance from country estates and a single important (capital) city. 'In Tudor times', writes Wiener (1981: 47), 'visiting Italians were struck by the absence outside the capital of the kind of urban life with which they were familiar.' From the late Middle Ages onwards, successful merchants would gravitate towards rural estates. These factors partially account for the special status that the countryside had achieved by the nineteenth century: 'Nowhere else is landscape so freighted as legacy. Nowhere else does the very term suggest not simply scenery and *genres de vie*, but quintessential national virtues' (Lowenthal 1991: 213).

The most crucial point about the nature–nation assemblage – its embodiment of identity and belonging – remains significant despite its neglect of one-fifth of all British citizens, the genuine country folk whose interests often contradicted the veneration of their domain (Miller 1995). Nonetheless, unlike the American admiration for untouched nature, for the vast majority of the British population, 'English culture tames and adorns nature.' The countryside's

> features are compages of datable cultural acts, mostly ascribable to ancestral precursors. The past that permeates this landscape is not the primordial wild, but a nearer history infused with memorable human processes, desires, decisions, and tastes.
>
> (Lowenthal 1991: 216)

Landscapes like the British countryside are both beautiful – counteracting the impact of agricultural modernity – and deeply meaningful as 'compelling symbols of national identity'. Patriotic feelings are invested 'in talismans of space and place: hills and rivers and woods become ideological sites of shrines, battles, birthplaces' (Lowenthal 2000: 198). Gradually diffused throughout British society, such visualised, tangible ideals provided a new framework for national identity that was steadily strengthened in the early twentieth century and galvanised by the world wars. Indeed, the success of rural imagery in sustaining morale at home and on the battlefield is testament to the identity-related quality of the countryside, working its magic across the entire political spectrum. The horrors of the First World War cast the humanised, gentle countryside into paradisiacal light, combining both nostalgia and utopia (Wiener 1981), while the Second World War prompted an 'outpouring of patriotic works on British land and life, environment and character' (Daniels 1993: 222).

Given the linkage between landscape as cultivated nature and national identity, it is not surprising that the countryside has achieved a privileged status in Britain. In terms of transatlantic comparison, it is '[t]he English iconography counterpart to American wilderness' (Coates 1998: 112). The crucial difference is the integral human element in British (and broader European) conceptions of the valued landscape – the 'middle landscape'. Consequently, the countryside ideal remains poorly developed in the US (Bunce 1994), while, across the Channel in France, it was alive and well and remains so today. But the core distinction was that a Frenchman's identity was 'rooted in the soil' of a productive, peasant-based agricultural landscape, which avoided the dominant British emphasis on visual garden-like qualities.

Nationalism and 'Cultures of Nature' across Europe

Compared to other European countries, France has been a pioneer in the field of landscape/nature protection. The Inspectorate of Historical Monuments, created in 1830, codified and extended a long-standing interest in historically symbolic monuments and buildings (Whited et al. 2005). In 1901, the *Société pour la Protection des Paysages* was founded – mirrored in England by the *National Trust* (1904) and in Germany by the *Bund Deutscher Heimatschutz* (1904). By then, the admiration for historical monuments had broadened to include consideration of natural monuments and associated landscapes, as expressed in France's ground-breaking 1906 'Law on the Protection of Natural Sites and Monuments'. The concept of heritage, now well-established, began to coalesce with the 'rediscovery of nature' (Walter 1990: 235) and yielded a specifically French version of the nature–nation assemblage. Few have described the new 'natural' and land-based patriotism better than the French parliamentarian Maurice Favre (quoted in Walter 1990: 235):

> Patriotism, Gentlemen, is not only a moral entity, an abstract conception, a geographic or historical expression. It is, in a certain way, the material and visible representation of the country itself, its particular physical character and its diverse elements, with its mountains, forests, plains, lakes, rivers and river banks, the varied and multiple aspects of its soil, such as have been transmitted by the slow succession of the centuries. Certainly, patriotism is an innate and instinctive sentiment, but that which contributes most toward fortifying it and engraining it indelibly in the peoples' souls is the attachment to their natural soil, to their preferred horizons, and the memories which this evokes.

The 'back-to-nature' mood that pervaded Europe by the late nineteenth century found its expression in two broadly related areas of anti-Enlightenment thought and identity-creation. The more aggressive nationalist variant drew on Romantic and conservative philosophies, whereas the socio-cultural variant sought to counteract the modernist transformation of people's lifeworlds – in their lived aesthetic, socio-economic, and spiritual experience. The former became an attractive option when a nation felt beleaguered by a dominant neighbour and thus figured prominently in the Franco-German relationship. Both countries 'idealized the eternal peasant, the harmony of a people, strong in their physique and in their morals, with their soil', exemplified by the 'moral order' regime after the Franco-Prussian war (1871), by Méliné's 'retour à la terre' (1896–1898) or the German slogan of 'Blut und Boden' in the 1920s (Walter 1990: 236). The latter went beyond the dislike of foreign domination to include a battle of ideas and essences, whereby quintessentially Germanic qualities had been 'debased' ever since the

Roman Empire, and more recently by the rationalist philosophy of the French Enlightenment (Coates 1998: 165). Historians seized on ancestral myths and exploited their presence in the public realm. In an analogy to the Turner thesis on the American frontier, Wilhelm Heinrich Riehl presented an 'environmentalist' explanation of the Germanic character: 'the primeval German forest had brought forth an equally primeval stock of hardy and resourceful Germans' (Imort 2005: 59).

Finnish history parallels, to some extent, the use of nature in German intellectual circles, reflecting a comparable context of nationalist self-determination and defence. Forests and watercourses were the central elements of an emerging national mythology since the early nineteenth century, rendered by painters into rural scenery and iconic national landscapes (Myllyntaus 2003). Overall, however, Finnish Romantic nationalism included the same duet of ethnic nationalist and socio-cultural 'Arcadianism'. While rapids and waterfalls were often used to illustrate the Finnish 'struggle against russification' and the dream of independence, rural scenes and forested wilderness equally symbolised a traditional way of life: a 'pure, unspoilt nature [... marking the] contrast to the decadence and complexity of modern industrialising civilization' (ibid.).

Norway was the only European nation that showed clear similarities with the American approach to nature/wilderness. On the one hand, Norwegian *Naturlyrikk* elevated raw nature to the state of spiritual inspiration, harnessing nature 'in the service of national sentiment' and making it an object of pride and admiration (Witoszek 1997: 214). On the other hand, like in the US, most Norwegians consistently displayed a utilitarian attitude to nature conservation. They believed that Norway had enough nature and should use its ample resources for economic progress (Whited et al. 2005). The impression of an American-style, bifurcationist humanity–nature relationship would nonetheless be premature, as the vast majority of Norwegians harboured rather pre-modern attitudes. Their roots lay in medieval pragmatism where '[h]uman life imitated natural rhythms' (Witoszek 1997: 220). The humanity–nature relationship was marked not only by continuous struggle but also by respect for and reliance on natural processes. Norway followed its European neighbours by enacting the 1910 Conservation Act which protected areas of special ecological and historical value.

While Finnish and Norwegian environmental histories contained elements of a wilderness-based approach, all other European nations gravitated towards the German and French models of landscape protection and identity-based nature–society dynamics. A leading influence on this transition was the German botanist Hugo Conwentz. His lectures were well received throughout central and northern Europe, leading governments in the Czech Republic, Denmark, and Sweden to consider the links between environmental features and national or regional identities: for instance, in the form of 'natural monuments' and productive yet aesthetic cultural landscapes. Conwentz integrated a variety of interests stemming from the natural sciences, public policy, and national cultural politics, always culminating in a critique of unrestrained industrial modernity and its deleterious effects on nature, traditional landscapes, and public morality (Whited et al. 2005: 189–191).

Overall, the 'politics of nature' in the nineteenth and early twentieth century can be regarded as a reaction to the advance of rationalist and economic modernity. The nationalist fervour that welled up during this period represented a double-edged sword. It not only provided a measure of symbolic and emotional compensation for unsettling changes but also made this continuing transformation possible by presenting itself as the saviour of faith, meaning, and belonging in the modern world. More significant for my argument – and its application to GMOs – is resistance at the level of everyday life and traditional lifestyles – that is, the socio-cultural variant. Eder's (1996) description of the 'double process of intensification' (of modern rationalism and cultural resistance) springs to mind. The 'back-to-nature' movement was not simply a project of 'Romanticisation' which tried to prevent the cultural disintegration of society. Not only were many of its objectives widely shared across the political spectrum and society as a whole, but they equally spoke to people's lived experiences of socio-cultural and environmental change – in the language of *Heimat* (homeland), heritage, pride, identity, and belonging – thus planting the seeds of a new moral and aesthetic attitude to nature (Rollins 2000).

The confluence of various elements from Europe's environmental history becomes evident here. Europe has long been a continent of hybrid landscapes which have given rise to the pastoral ideal combining aesthetics, natural order/morality, and agricultural production. In most European countries, this classical, medieval (and possibly postmodern) conception represents the baseline for the cultural politics of conservationism. The case of Germany provides a pertinent example for the continuing relevance of this integrated framework, its fusion with the nationalist mood, and its resistance to processes of rapid modernisation. 'Re-enchanting' historical monuments and even less remarkable landscape features (such as groves, ecological habitats, rivers, and moors), nature preservation, and homeland protection (*Heimatschutz*) joined together in a common effort to protect and sustain the 'middle landscape' – the quintessentially cultural landscape (Lekan 2003). Rather than merely being a reactionary movement hooked on aesthetic nostalgia,

> [t]he concept of *Heimat*, [...] a signifier of regional diversity and a sense of place, became an aesthetically charged and socially conservative pathway to broader ecological understanding. [...] The idea of the cultural landscape offered a precarious middle ground, rather than a sharp dichotomy, between wilderness and civilization within Germany's rapidly industrializing and urbanizing society.
>
> (Lekan and Zeller 2005: 4)

These developments were not exclusive to Germany, but could be observed in most European states, to varying degrees and at different speeds.

From cultural landscapes to national parks

If this 'via media' is represented by Europe's interactive humanity–nature relationship, then it should also be prominent in thought and practice regarding national parks and cultural landscapes. For a long time, conservation strategies across Europe focused on scattered areas of outstanding historical value or natural beauty, enabling a surge in leisure tourism. The earliest challenge to this tendency could be found in Britain where the conservationist movement's concern for traditional architecture and landscapes had many powerful backers in society and in Denmark, which implemented a concerted plan of landscape protection. However, most European conservationists were convinced that it was futile to fight against the industrial juggernaut, choosing instead a strategy of accommodation and salvage of individual areas of special value. Such fatalist pragmatism tallied with the attitudes of American preservationist movements and their strategy of saving 'sacred' places while also enjoying the fruits of 'progress'. At a superficial level, this indicates a form of transatlantic convergence because Europe in the early twentieth century appeared to catch up with the American wilderness tradition, exemplified by a wave of newly created national parks in Sweden (1909), Spain (1918), Italy (1919), and France (1927). According to this view, European conservationists were engaged in a belated battle to save the remnants of wilderness in their respective countries. Yet, this perception was scarcely realistic, given Europe's cultivated and hybrid environments – with the possible exception of Alpine ranges and wind-swept seascapes.

In 1976, the American historian Roderick Nash delivered a series of lectures in Italy and concluded that, despite the network of Italian national parks, 'there is no wilderness left to preserve' (quoted in Hall 1998: 24). By contrast, Hall points to the unfamiliarity of wilderness in the Italian context and suggests that it refers to less spectacular 'natural' features, such as 'brambled woods, and the occasional *stambecco* or *camoscio*', that is, furred game. Rather than worrying about the absence of 'true' wilderness, Italian professionals were engaged in the task of ecological restoration: 'the conversion of damaged areas into ideal landscapes' through reforestation, remediation, landscape planning, and maintenance (Hall 1998: 27). The American variety of wilderness thinking, increasingly influential since the early twentieth century, relied on a mixture of cultural power and scientific legitimacy. Imbibed by foreign tourists visiting US national parks and beamed across the globe by television broadcasts, it offered an image of spectacular and spiritual nature, of genuine otherness and wonder. Early ecological science was used by a scientific elite to decide on the balance between economic development and conservation, on what had real 'value' and what could be sacrificed.

Post-war West Germany provides an especially pertinent example of how scientific perspectives were employed to neutralise identity politics which had played such a disastrous role under the Nazi regime. The *Heimat* tradition, which emphasised emotional and socio-economic connections to the land, was largely abandoned in favour of 'scientific ecology', holding out the promise of a 'presumably more objective and neutral program of environmental protection and restoration' (Lekan 2003). It is testament to the resilience of land-based patriotism that *Heimat* remained a popular way of overcoming the consequences of military defeat and affirming a positive identity rooted in natural and cultural heritage, especially at local and regional levels (Chaney 2005). In East Germany as well, *Heimat* still loomed large in official discourse, albeit articulated with a distinctly socialist slant.

Many of today's conservationists look back with dismay to a period when 'we have denigrated folklore in our preference for scientific rationalism and we have consequently lost our associated landscapes' (Fowler 2001: 72). In Europe, insists Phillips (2001: 52), nature and culture remain closely interwoven to the point where 'cultural values cannot be measured only in terms of great monuments but also exist in the spiritual significance and cultural importance attached to landscapes by human communities.' The resurgence of cultural landscapes is a European-wide (and probably global) trend. European landscape planners and conservationists have increasingly tapped into a prevalent 'common sense' in which the choice is rarely between a worship of pristine nature and corrupted wastelands. Rather, popular attitudes frequently rely on a sense of 'long term dwellingness of place' (Macnaghten and Urry 1998: 273), where local/regional histories, environmental stewardship, and 'environmental identities' encourage 'homeland' pride and defensiveness against perceived disturbances and economic commodification.

Because it is difficult to neatly separate aesthetics from other elements, the concept of identity best captures the spiritual and socio-economic aspects at the heart of cultural landscapes. Both the Convention Concerning the Protection of the World Cultural and Natural Heritage and the European Landscape Convention recognise the long history of interactive humanity–nature relations, acknowledging that '[a]t the core of the landscape concept is the spiritual relationship between people and nature' (Monteiro 2005: 10071). Cultural landscapes are commonly regarded as

> natural settings, engraved with ancestral use, the ongoing, modern-day use of which preserves a piece of history and reminds us of our heritage. This observable ancestry gives the natural setting its identity, and with an identity, a space becomes a place.
>
> (ibid.)

Such intensely perceived cultural identity, engraved into the physical environment, is arguably the overarching function of cultural landscapes. Its relationship to national identities has already been explored above.

Many scholars in the field of conservation history agree that a sense of the ideal past and future as well of places of dwelling and belonging are at the core of people's emotional connection to landscape and nature (Fowler 2001; Hall 2001). Of course, economic activities are also integral to this vision, especially food production through agriculture. In this sense, the alternative to the 'ecosystem approach' of early conservationism is not merely the touristic version of aesthetic delight and outdoor recreation, but the 'agri-cultural' landscape – an integrated whole of ecological maintenance, livelihoods, and cultural identity (Monteiro 2005).

European policies on landscape conservation reflect the multiple rationales behind landscape protection. Pioneering countries – among them Britain, Austria, Germany, France – trace their first laws back to the nineteenth century when the preservation of national heritage was still the main priority. Countries such as Italy, Norway, and Sweden followed suit in the early twentieth century. Post-war Europe witnessed the ascent of ecological science and conservationist arguments, but the long-standing divide between city and countryside now turned into something of a lifeline for pastoralism, as urbanites rediscovered their taste for rustic country life and 'natural' landscapes, giving rise to a boom in tourism and aesthetic appreciation. From the late 1980s onwards, and spilling over into some central and eastern European countries after 1990, the productivist model of agriculture came under increasing pressure from ecology, tourism, and market constraints. Austria's agri-environmental support framework for Alpine farmers and Norway's targeted production subsidies for farmers in marginal areas illustrate the new integrated thinking especially well, and so do other measures that protect geographical labels of origin (PGI certification) or create markets for organic foods. This trend will be further explored in Chapter 6.

Besides a long-standing concern for landscape protection, the character of Europe's national parks underlines the disparity with the US. At first glance, the surge in the number of national parks over the last century seems to corroborate a growing infatuation with the wild. Italy, for instance, created four national parks within little more than a decade: Abruzzo and Gran Paradiso in 1922, the Pontine Marshes in 1934, and the Stelvio National Park shortly after (Sievert 1998). By 1990, national parks covered 4.3% of national territory and doubled again to roughly 10% by the year 1998 – significantly more than other countries. Yet, the reality of nature protection in Italy is peculiar. As conservationist Roberto Gambino (1998: 31) explains, 'Italian parks and their surrounding areas are characterized by an exceptional richness of historic and cultural heritage and a high intensity of anthropic land uses.' Even famous parks like Abruzzo are usually composed of private lands with only a fraction of public ownership.

In France, this picture of mixed protected landscapes is largely being mirrored. Here, 'natural' areas (nature reserves and national parks) cover around 8% of national territory, of which only around a quarter is strictly protected from human activities. Another category (Parcs Naturels Régionaux) covers around 11% of national territory and is subject to fairly lax regulations, reflecting their composition of 'woodlands, fields, villages, country lanes, meadows, and farms' (Bess 2003: 222). The broader tendency is towards 'middle landscapes', even when it comes to protected spaces. Northern European countries largely bear out this picture. In Britain, national parks (covering 8% of national territory), heritage coasts, and 'Areas of Outstanding Natural Beauty' (15% of territory) have increasingly been informed by 'a sense of inter-dependence between natural beauty, countryside recreation and the well being of rural residents' (Dower 2001: 102).

Germany did not designate any official national parks before the creation of the Bavarian Forest National Park in 1970. Nature reserves had existed for some time, but it is revealing that the first reserve, the 1921 Lüneburger Heath in northern Germany,

> acquired its characteristic vegetation due to agricultural usage, rather than ecological succession alone, while man-made landscapes – rural hedgerows, rock quarries, railway corridors – provided the ecological niches for birds and other animals that environmentalists now see as 'natural', indigenous species.
>
> (Lekan and Zeller 2005: 5)

Once the post-war conservationist momentum had begun, nature parks mushroomed in West Germany and covered 15% of national territory by the early 1970s. There was equal enthusiasm in East Germany. Yet, as in other European countries, few areas within these reserves were stringently protected and most funding was allocated to 'recreational facilities' (Chaney 2005).

Europe's experience with nature conservation, park creation, and mixed land use is, of course, not uniform across different countries. Denmark, for instance, did not decide to single out areas for heightened protection, preferring instead to focus on the restoration of habitats and the protection of rural landscapes. This only changed in the wake of the 1979 EU Birds and Habitats Directives when specially protected spaces became mandatory (Dower 2001). Thus, while there is no harmonised European landscape and conservation legislation, there is enough similarity to assume an underlying civilisational resemblance. Even before the advent of supranational regulations, European laws displayed significant commonalities in international comparisons (Monteiro 2005), grounded in the age-old interactions between humans and their environments and consequent emotional bonds. This condition often makes it possible for human interventions to be perceived as 'natural' elements within an essentially cultural landscape. The discourse of conservation thus assumes a broader social meaning: the protection and reproduction of the lifeworld, cultural practices, identities, and 'natural' surroundings.

Conclusion: The bifurcated versus the 'middle' landscape

The major argument of this chapter is that Americans have more convincingly performed the separation of the human and the natural spheres: relative bifurcation in the US can be juxtaposed with a more interactive tradition in Europe. A combination of material and ideational elements provides a plausible civilisational account of why landscape and prevalent environmental attitudes in America are distinctive. In the European case, the core civilisational dynamic is the intermingling of humanity and nature, both in ideational terms and for material reasons of sustaining livelihoods under conditions of relative scarcity. As I will argue in the final chapter, the affective dimension of this relationship it is also particularly relevant for understanding the cultural politics of agbiotech in Europe. Conversely, America's bifurcationist heritage is a major underlying reason for why the anxieties roused by agbiotech in the US have largely been confined to utilitarian considerations – as well as, more recently, to individualist values of 'consumer sovereignty'.

America's bifurcated landscape is both a landscape of the mind and a material landscape. The great majority of Americans accept that large parts of the country are reserved for 'material', economic needs, while a much smaller proportion is devoted to satisfying emotional and spiritual needs. The 'middle' landscape of pastoralism has progressively disappeared and farming has become a business like any other (Worster 1994b). The Jeffersonian rural ideal has been substituted by middle-class suburbs (Tuan 1974). Transcendentalist Romantics reacted against rapid industrialisation by re-spiritualising nature, but they did not challenge the underlying trajectory. American Romanticism, though still a powerful cultural force, has not been able to break down the historical boundaries between humanity and nature (Coates 1998: 127). Romantic infatuation with the sublime has contributed to the tainting of the middle landscape as mediocre and the urban setting as abhorrent.

When America entered the twentieth century, the heirs of Romanticism and supporters of capitalist economic progress were busy carving up the land according to their respective principles. Although I do not wish to downplay the intensity of the struggle between preservationists and 'wise-use' conservationist-utilitarians, both groups accepted the need for a fundamental separation in land use. Their quarrel was about which specific areas should be reserved for wilderness. Later on, when the commercialisation of nature had advanced sufficiently, the contenders would also be united by economic motives. National parks have turned into genuine tourism magnets which offer active recreation without forgoing the pleasure of hotels, cabins, and visitor centres. Along the US East coast, the region of the first settlements, there still exist alternative interpretations of nature reserves that are rooted in a much more European understanding. The Adirondack State Park, north of New York, is the largest state park in the US and features extensive human use and occupation, including 130,000 inhabitants (Beinart and Coates 1995). The accommodation of a human presence mirrors European nature reserves which almost always rely on such compromises. However, it took an extraordinary 106 years of environmentalist campaigning for Adirondack to achieve its present status (Bray 1998), and it remains a *state* rather than *national* park.

Whereas, in the American West, native populations could be removed to make way for 'nature', the East coast has been characterised by more settled communities and patterns of land use, making it less susceptible to the dualism of wilderness and civilisation. Contemporary Eastern landscapes are a reminder of a bygone age and a much gentler variation on the theme of bifurcation. Nonetheless, the overall spatial and conceptual differentiation appears to be locked into a deeper American civilisational structure and is reproduced by the contemporary 'culture of nature'.

> Perhaps only an American cultural perspective hooked on wilderness as the supreme manifestation of nature, and shaped by the comparative recency of the large-scale transformation of the natural world in North America, has to work so hard to grasp the absence of clear boundaries between nature and culture.
>
> (Coates 1998: 234–235)

Nonetheless, theorising about Protestant reform, the ramifications of capitalism, and aesthetic reactions might suggest a historical transformation of the entire Western world (and probably beyond), a point underscored by Eder's (1996) 'double process of modernity'. Although there is some truth in this assumption, my historical summary of American 'environmental identities' has emphasised that Americans can lay claim to a unique cultural evolution.

Modern rationalism and industrialisation have, to a large extent, also superimposed themselves onto European societies. And there were influential sections of European societies, as well as the state apparatus, that actively encouraged and financed the transformations wrought by the industrial and agricultural revolutions. But, as the example of pioneering Britain shows, significant parts of these same elites would lose their zeal and become converts of the spirit and beauty of traditional, cultural landscapes. 'The English character', deplores Wiener (1981: 6), 'was not naturally progressive, but conservative; its greatest task – and achievement – lay in taming and "civilizing" the dangerous engines of progress it had unwittingly unleashed'. This remark, further substantiated in the final chapter, broadly applies to all European societies. Unlike in America's Protestant colonies, there has never been a radical break with the past in Europe and, to some extent, this has allowed established practices and patterns of thought to persist. Both intellectual and popular reactions to modern transformations have sought to rehabilitate elements from the 'golden past', whereas Americans were often beguiled by dreams of a luminous future.

Paraphrasing Baudrillard (1988), Macnaghten and Urry (1998: 184) describe American landscapes as paragons of pastless, individualistic, meritocratic modernity: 'American landscapes are empty and as such they stand for modernity and the rejection of the countrysides of Europe and the complex histories of European societies. Such an emptiness is a metaphor of the American dream.' In Europe, cultural landscapes – combining elements of familiarity, spiritual or cultural meaning, aesthetics, and livelihood – represent the humanised environments inhabited and admired by the vast majority of citizens. Often identified as traditional and 'natural' in common parlance, they are best identified as landscapes of dwelling and belonging. A focus on ecological or aesthetic qualities of landscapes would only imperfectly describe the irrepressible appeal of cultural landscapes. As far as the link between nature and human identities is concerned, the rehabilitation of cultural landscapes is an important consideration and sets the stage for a more detailed exploration of public attitudes towards GM crops and foods in the final chapter. In Europe, the intermingling of nature and humanity – so characteristic of the ancient world and the Middle Ages – is staging a late-modern comeback.

6
Agri-Cultural and Culinary Identities

Introduction

Chapters 3 and 5 suggested that defensive reactions to modernity's rapid environmental and socio-cultural changes can often be regarded as instances of identity politics. Nostalgic visions of the past and images of nature or the ideal landscape are employed as symbolic 'ramparts' against the forces of rationalisation and industrialisation. The concept of nature also figures prominently in such counter-movements. The present chapter builds on these insights and explores their relevance for the political controversy over GMOs. It concentrates on the analysis of European developments because, despite a recent upsurge in activism, a comparable resistance to agbiotech has not yet developed in the US. The previous chapter's historical overview also underscores another fundamental point. As Schaffer's (1997: 124) quotation (in Chapter 3) implied, early modern and late modern resistance to 'progress' has important parallels, particularly the tendency to fuse notions of cultural and natural order and insist on the centrality of customs and 'heritage'.

There are many European examples of the defence of specific nature-identity assemblages, often discussed under the label of NIMBY-ism – a 'not in my backyard' resistance to changes in local landscapes and amenities. Wind farms producing renewable energy might be regarded as desirable developments, but opposition to them has been growing throughout Europe. Recalling the values embedded in familiar, naturalised and cultural landscapes, it is not surprising that such projects pose a challenge to local 'environmental identities'. For instance, in early 2008, a major wind farm proposed for the Scottish Isle of Lewis was rejected due to overwhelming opposition from islanders. One resident explained her resistance in terms that illustrate the locals' fundamental connectedness to their surrounding environments:

> We've been brought up to respect and love the moor ever since we were tiny. [...] It's a piece of ground which means so much to us. It's just part of us... I knew right away that I didn't want the moor to be dug up or concreted over in any way. That would just break my heart. I don't see how it will be anything other than a completely devastated area.
>
> (Carrell 2008)

This is not necessarily an irrational reaction. Psychologically speaking, drastic environmental change, ranging from industrial development to climate change, can disrupt a stable sense of the self. Albrecht et al. (2007) call this condition 'solastalgia', derived from *solacium* (solace) and *algos* (pain), concisely explained as 'the homesickness you feel when you're still at home' (Albrecht quoted in Skatssoon 2005). Barry et al. (2008) highlight the need to respect local people's emotions and their underlying values while searching for common ground. And Lowenthal (2000: 198) acknowledges that NIMBY opposition has often been a 'rational', if 'deeply parochial' response to the threat of transformation. The identity-based character of local opposition becomes evident if one considers that – in the case of Lewis – other small-scale, artisanal practices, such as crofting, appear to be an admissible way of 'digging up' the moor. Cumulatively, the gradual decline of crofting and age-old links with both lifeworld and landscape makes it an unlikely target for similar public mobilisation. It is part of the customary *interaction* with nature and the moral economy, and not integral to its perceived industrialisation.

In this final chapter, I develop the idea that the transatlantic divergence over agbiotech is rooted in a broadly similar – albeit much more abstract and less localised – cultural politics of nature. Divergent concepts of nature, as well as the values and identities associated with them, are powerful (but indirect) drivers of the contemporary EU regulatory system. Nowhere is the linkage between cultural identity, images of nature and anti-GMO attitudes stronger than in the areas of agriculture and food consumption. These have been dissociated from nature more successfully in America – in both material and ideational terms. To many Europeans, GMOs are frequently perceived as harbingers of a 'post-natural' or 'post-human' future, and appear as 'genes out of place', and as disturbers of settled and cohesive identities. They crystallise a suspect, instrumentalist approach to the world (Marris et al. 2001: 69) and serve 'as a sounding board, or a projection screen, for deep differences in interests and world-views' (Torgersen et al. 2002: 23). In the following, I elaborate on the theme of 'disenchantment' and develop the idea that anti-GMO mobilisation is part of a long historical trend of reactive responses to rationalist modernity. Second, I return to the inextricable assemblage of nature, culture and identity which I illustrated in the previous chapter, and I suggest that the values and landscapes associated with agriculture are perceived to be under threat from agbiotech – which is seen as one of the latest incarnations of modernity. The cultural power of these attitudes is emphasised by tracing their influence on national and European policy-making. Finally, I apply similar reflections to the topic of food cultures.

Disenchantment and colonisation: A long historical struggle

If pro-agbiotech actors seek to truly understand their opponents, they will have to pay more attention to cultural forms of knowledge. Just as wind power companies cannot simply rely on their 'green' credentials, agbiotech supporters would have to cease presenting their inventions as 'natural' and self-evidently beneficial, while branding their detractors as ignorant Luddites or economic protectionists (Gaskell et al. 2001). As the biotechnology coalition seeks to establish its economic and societal presence, it has to cope with a pre-existing cultural environment in which its own preferred discourses may not always resonate adequately (O'Mahony and Skillington 1999). The relevant context in European countries is deeply influenced by two long-standing socio-cultural processes which previous chapters have illustrated in a variety of ways: the disenchantment of nature and the rationalist, utilitarian reshaping of the lifeworld. In the following, I expand on the significance of these two conditions, thus setting the scene for further analysis of agriculture and food cultures.

Rationalism and the evacuation of meaning

First theorised by Max Weber (1930), the disenchantment of the world refers to the decline of religious and other supernatural worldviews during the seventeenth and eighteenth centuries that had hitherto legitimated a cosmological order and filled the natural world with deep spiritual meaning, while also assigning a specific role to humankind. Disenchantment expresses

> a steady decline in the portrayal of the physical environment as possessing venerable supernatural, spiritual, or organic qualities, and a growing tendency to regard the earth and its creatures as being simply

> forms of mechanistically structured matter, subject to empirically ascertainable causal laws, and available for humans to manipulate and put to use as they saw fit.
>
> (Bess 2003: 117)

Translated into an identity-based framework, the waning of pagan and Christian beliefs also undermined some aspects of people's identities, leaving precious few markers about the origin or purpose of human existence and casting doubt on the idea of a stable 'natural order'. It is, however, easy to overestimate the transformative effect of this process on persistent cultural beliefs and everyday practices. As I argued in Chapter 3, religious or otherwise moral conceptions of natural order still underpin the contemporary 'politics of nature'. As late as the end of the nineteenth century, in many sections of society these ideas were highly explicit rather than submerged: for example, '[t]he conviction that animals and vegetation had religious or symbolic meaning for men remained an article of faith for many Victorian country folk, [...even if] it no longer had the support of intellectuals' (Thomas 1984: 91).

Today, the idea of 'naturalness' has made a comeback, but this resurgence is not a straightforward reaction to disenchantment. First, a similar process of 'desacralisation' applies not only to the realm of nature but also to social institutions themselves. Scientific and bureaucratic rationalisation served to demystify and reorganise institutions by turning them into calculable, 'rational' services, often subject to external oversight and strict rules of efficiency – Weber's 'iron cage' of bureaucracy. Examples of such 'social machineries' are modern state agencies, medical institutions, and large corporations. Second, this shift implied more than a replacement of traditional authority by natural science, which is the prime source of legitimacy in the modern age. The main problem is that science by itself does not contain a moral faculty which could put forward a vision of the good life (Heller 1986). Nor does it allow for the intersubjective reproduction of customs and meaning that are integral to everyday life:

> Science doesn't imply a lifeworld; it also draws knowledge away from everyday life into the specialised spheres of reproducible know-how, only returning it to everyday life in the form of rationalised and mechanical practice. In these transformations everyday life may well seem to lose its adequacy and necessity as experience.
>
> (Wright 1985: 18)

In other words, the modern condition is characterised by the destabilisation of everyday life, of familiar or traditional socio-cultural practices and identities – a process which Habermas (1984) called the 'colonisation of the lifeworld'. Habermas essentially updates Weber's theory of rationalisation and synthesises a number of concepts drawn from other social theorists. He posits that human well-being and social stability rely on a balance between a lifeworld of communicative-intersubjective relationships and a highly rationalised, efficient economic system that ensures physical survival and generates wealth. With regard to the cultural politics of nature, NIMBY-ism, 'solastalgia', and political and cultural resistance are prevalent reactions to this process of 'colonisation'.

The late modern response

The juxtaposition of lifeworld and system has inspired many analyses of contemporary social movements, including perspectives on anti-GMO coalitions. Zwick (1998: 86) thus describes biotechnology as a socio-economic modernisation that is initially evaluated on the basis of technical knowledge, efficiency, competitiveness and profit. However, it is then pitted against a public appraisal based on quality of life, social and environmental compatibility, and ethical correctness. Technocratic modernisation opens up a gap between system and lifeworld and the latter fights back with powerful images of nature and human well-being. A 1972 survey, in which about 50% of French citizens considered that the costs of modern technological progress outweighed its benefits, provides an early indication of such broad-based societal disillusionment (Bess 2003: 87). Eder (1996) theorises a dual dynamic of intensification: the stronger and more irresistible modernisation became, the more powerful grew the appeal of its identity-based counterpart, expressed in a yearning for nature, tradition and a golden past.

The separation of tasks between system and lifeworld held up sufficiently well for over a century. In both America and Europe, nationalist ideologies – fortified by associations with nature and heritage – allayed the effects of disenchantment by fostering renewed feelings of sacredness and belonging. A seemingly endless wave of 'pastoralising adverts' in Europe (Wright 1985: 19) and wilderness iconography in America attest to this linkage. Yet, the ideologies of the Enlightenment that dominated the nineteenth and twentieth centuries – liberalism and socialism – have never successfully tackled the question of identity (Eder 1996: 191). As the twentieth century progressed, the concept of nature was increasingly used to project the associations I catalogued in

Chapter 3: natural order and a sense of spirituality, purity and authenticity, goodness (of essence and intentions) and stability. Picturesque landscapes, national parks and monuments were frequently called upon to embody these virtues.

For decades, if not centuries, nature and the 'golden past' have acted as a counterbalance to the 'system' and continue to do so today by offering a measure of 're-enchantment'. As Wright (1985: 21) puts it:

> [W]hile everyday life is more and more dominated by the repetitive thinking which is demanded by various rationalised and routinised activities, the more intuitive and analogical modes of thought don't simply cease to exist – even though they too may wander off to imagine the signatures of natural meaning in the past and countryside.

Towards the end of the twentieth century, the implicit compromise that fuelled the 'double process' of intensification became unstable. For Skillington (1999: 190), we have reached the apogee of an ominous historical process: 'From the nineteenth century onwards, there has been an improved tuning of the behaviour of both humans, animals and plant species to the rhythms of the machine as the "destruction of nature" has become institutionally standardised and integrated into the workings of everyday life.' Nature and machine are no longer clearly distinct from each other. The old opposition is dissolving, for plants and animals can now become organic machines themselves, working tirelessly for their human masters. Biology itself ceases to be a solid foundation of meaning (McKibben 2003). This 'spiritual resentment at the loss of meaning in a nature enframed by technology' (Scott 2000: 304), and the underlying critiques of modern 'progress' and cultural change, represent public sentiments that are difficult to uncover through opinion surveys. The charge of 'unnaturalness', which I identified as the guiding theme in both quantitative and qualitative studies, represents the top of a vast interpretive 'iceberg' of deep cultural structures.

Nature and the ideal past in America and Europe

By contrast, when it comes to technological innovations such as GMOs, a significant majority of American citizens see neither nature nor the idealised past as threatened with oblivion. In fact, many continue to regard nature as a threat – to agricultural productivity, human security, and economic fortune. Even in a growing positive sense, which Nash (1967) has chronicled, nature was largely dissociated from everyday life.

Turner's 'frontier thesis' expressed the spirit of the age by theorising the making of a new human being thanks to the wilderness condition. Another significant role of nature was a moralistic, Republican version of the concept which expressed the Enlightenment's faith in natural law and scientific truth as the governing mechanism of the universe. Once more, nature appeared as a thoroughly modern force, as a divider rather than bridge to the past.

Later on, by the end of the nineteenth century, the 'double process' of intensification would take its course in America as well. As a remedy against materialist greed and scientific rationalism, nature as wilderness could partly fill the void of spiritual objects and create some transcendental meaning in a world of industrial capitalism (Vogel 2001). Moreover, enshrined in national parks it could now function in a similar way to cherished European landscapes, as a manifestation of national identity and signifier of American heritage, of the wild purity that allegedly characterised the New World at the time of the first settlements (Coates 1998: 108). The Enlightenment variant of nature as universal truth, however, has retained its strong legitimacy in the US and remains largely unassailable both in public policy and popular culture. In contrast to the climate change debate where the US public is uneasy about controversial decisions being based on (allegedly) contested scientific evidence, in the case of agbiotech, the belief that current scientific knowledge is 'yielding universal and objective facts that provide the basis for impartial regulatory decisions' (Skogstad 2006: 220) is a key reason for relatively permissive regulations. Along the lines of comparative sociology, in Europe cultural and political factors more frequently trump the technocratic vision of universal 'truth' (Jasanoff 2005).

In summary, the 'environmental identities' of most Americans are closely wedded to a spectacular landscape that remains largely dissociated from everyday life. When 'nature as nostalgia', associated with national heritage and an idealised past, comes into play, it sets up an impossibly pristine ecological idyll or draws on the virtues of the rugged and ambitious pioneer who subdues an uncivilised terrain. In the face of late modernity's technological innovations, the former can be set aside and preserved, the latter updated and celebrated. One finding from Peters et al.'s (2007: 205) survey on culture and agbiotech illustrates the persistence of the latter aspect: significantly more Americans than Germans agreed with the statement that 'humans are smarter than nature'. In Europe, the icon of nature and the 'aura of the modern past' (Wright 1985) contain the seeds of resistance. In most countries, especially Britain, a hegemonic, polished and largely consensual version of the past predominates. In the case of France, the national past has been purged of its divisive elements and reduced to family values, conviviality and the love of the countryside (Hoyau 1980). Apart from such biased reconstructions of the past, Wright (1985: 254) also highlights its ability to demonstrate the presence of 'radical social needs':

> The aura of the modern past can indeed speak against rationalised experience; it can celebrate a contemplative rather than instrumental or pragmatic relation to nature, just as it can pose the question of an unalienated rationality and of a social experience in which to be human is to be a personality as a whole rather than the subordinate or merely specialised subject of so much modern activity.

The past can therefore become a plea for an 'other' modernity that still embodies some of the original promises, such as quality of life or human health, but lacks the dark side of excessive rationalisation, alienation, and exploitation of nature. For most Europeans, agbiotech is associated with the latter elements. Yet, in northern European countries, one rarely hears an invocation of heritage or national identity in this regard. Much more often, the opposition is articulated through images of nature as well as agriculture or food. It is best to think of such north-south divergence as differences overlaying a fundamental civilisational similarity. As previous chapters illustrated, an interactive relationship between nature and humanity has a long historical pedigree in all European societies, but different countries have developed their own specific 'cultures of nature' (Herrick 2005). Agriculture and food cultures, in particular, highlight the ways in which the struggle to preserve and enhance cultural identities has become a central part of the campaign against GMOs.

Europe as an 'agri-cultural' civilisation

Agriculture, with its associated livelihoods, landscapes, and practices of food production, is a principal illustration not only of greater 'mixity' and overlap in land use patterns compared to the US (Peters et al. 2007), but also of identity-related aspects. Because many European societies continue to emulate the agrarian nations they once embodied (Zurek 2007), they may be contrasted with a more technology-friendly and utilitarian US that has left Thomas Jefferson's dreams far behind (Thompson 2001). The agricultural revolution, as well as its differential speed and impact across European countries, has been reviewed in the previous chapter. And this cascade of technological improvements and productivity gains is very much an ongoing process. The first image of Eder's (1996) 'double process' of intensification is alive and well in the twenty-first century, but so is the flipside of contemplation and 're-humanisation'. The philosophy of productivism, which takes the modern factory as its blueprint, still finds itself opposed to the second image and a diverse array of societal groups. Finally, it is important to counter the view that agriculture is of minor significance in northern Europe, and that the 'agri-cultural' perspective only applies to the south. While it is true that northern European attitudes and policies are more focused on negative environmental externalities of farming and on aesthetic-visual qualities, this does not signify the irrelevance of an 'agri-cultural' way of life, especially at a time of growing enthusiasm for local food and cuisine.

Agriculture and identity in northern Europe

In many European countries, agricultural modernisation only reached a breakthrough after the Second World War. The enclosures had entrenched a patchwork landscape of fields and hedgerows which the British, despite lamenting the loss of the commons, soon adopted as their ideal and 'natural' landscape. When this cultural landscape began to be threatened by economies of scale, land consolidation and bold modernist projects, a powerful counter-movement took shape, including old-style conservationists, NIMBY groups, aesthetic elites and environmental movements. Their sympathisers may well encompass the great majority of the population if one extrapolates from the frequency of supportive news coverage and membership numbers of prominent mainstream preservation organisations – such as the National Trust, the Royal Society for the Protection of Birds (RSPB), the Campaign for the Protection of Rural England (CPRE), and countless local amenity societies.[1] The RSPB and the CPRE, in particular, explicitly recognise the intertwined character of agriculture, aesthetics and biodiversity.

Such an integrated conception also counteracts charges of exclusive 'aestheticisation' of the British countryside. Nor are preservationists and their sympathisers all obsessed with the urban–rural divide. The encroachment of suburban housing projects on green fields is a common theme, but so is the elimination of green spaces within or 'green belts' around existing cities. There are, insists Herrick (Herrick 2005: 291), 'strong ideological and physical linkages between rural and urban' in European countries, including Britain where active farming is widespread and public access is ensured by a nationwide network of footpaths. Undoubtedly, the rural is sometimes idealised to the point where it crowds out positive depictions of the city and other mixed spaces, but this tendency is closely related to the trends of depopulation and mechanisation. National and increasingly global economic forces have effectively drained the countryside of its working population, turning peasant communities into agglomerations of self-styled 'agri-businesses'. According to Rowe (2001: 123), the farmers' symbiotic relationship with the land has been replaced by a technological 'way of life' whose purpose is to overcome nature and extract material value by transforming the 'traditional landscape fabric'. Rowe also points to another trend which reflects Eder's 'double process' of intensification: 'the public's appetite for the so-called "traditional countryside", with all its diversity and character, appears to be growing, [while] the forces of economic and technological change are moving in the opposite direction' (ibid.).

In a comprehensive survey from 1955, the English regarded their country as an essentially 'rural' nation (Wiener 1981: 73). Subsequent decades might one day be seen as a modernist aberration in what remains a strongly 'agri-cultural' society that views the countryside as a soothing natural landscape, a repository of national meaning. Conservationists with ecological inclinations also tend to condemn rationalisation and industrialisation, acutely aware that the protection of wildlife depends on a moderated form of agriculture. The government's own advisory organisation, English Nature (suitably renamed Natural England in 2006), warned that herbicide-tolerant GM crops would lead to a decrease of biodiversity and thus continue existing negative trends (Lang and Heasman 2004: 181). The world's biggest study on the ecological impact of GMOs in the early 2000s (the UK 'field-scale evaluation trials') largely confirmed these concerns. The benefits of spraying fewer pesticides were overshadowed by their more powerful impact on many insect species, leading to a decrease in the number of butterflies, bees, and ultimately farmland birds. The positive externality for human health of using fewer chemicals was potentially outweighed by the negative externality of 'less nature' in 'agricultural' landscapes. Against the narratives of ecological modernisation favoured by pro-GMO commentators, the new seeds were re-inscribed into the story of agricultural intensification and spiritual and aesthetic decline.

Britain is not the only northern European country that has combined the character of a 'working' agri-cultural space with images of the 'natural' and other identity-related themes. Germany, for instance, celebrated the imagery of rustic, virtuous peasant life in the first half of the twentieth century. Agricultural modernisation was an ongoing process for decades and reached its apogee between 1950–1970 when the *Flurbereinigung* (land consolidation) re-organised vast swathes of the West German countryside. The damming of rivers, building of roads, drainage of swamps, and creation of monocultures featured equally prominently in East Germany from the 1960s onwards, where it was facilitated by forced collectivisation (Whited et al. 2005). Ideologically tainted through its association with racial and nationalist discourses and marked by significant regional diversity, the German countryside has not been able to provide an equally solid foundation of national identity. But the Europe-wide revival of cultural landscapes has been palpable here, too. Today's federal and state-level ministers of agriculture will freely extol the virtues of productive, yet intensely historical landscapes and sometimes profess the goal of supporting 'peasant' agriculture as a model for the future.

Neighbouring Denmark has long pursued an integrated, traditional landscape strategy that also ensured economic viability. So strong was the affection for the 'agri-cultural' landscape that nature reserves were only created in the early 1970s to satisfy EU legislation, while naturally occurring afforestation was frequently viewed as spoiling the countryside's scenic beauty (Whited et al. 2005: 192). A final example from central Europe, the Czech Republic, shows that even formerly Communist states have often reverted to type. The 1992 Nature and Landscape Protection Act responded to the population's 'emotional attachment' to locally specific landscapes. Farming has faced greater scrutiny in terms of environmental impact, but equal scepticism has been cast on the 'rising tide of capitalism' (Dower 2001: 100) – themes consistent with both agri-cultural landscapes and the appeal of the 'natural'.

Agriculture and identity in southern Europe

Southern Europe, with the partial exception of relative 'outliers' such as Spain and Portugal, offers an even clearer picture of 'agri-cultural' sensibilities. As described in the previous chapter, the modern age had seen another rapid increase in environmental degradation in the Mediterranean and this also had socio-economic and cultural consequences. Momentous changes, for example, occurred in Tuscany, Italy, where deforestation and greater agricultural specialisation caused landscape diversity to decrease by more than 70% between 1832 and 2000 (Agnoletti 2003). Across the Mediterranean, the sale of previously communal or Church lands to new 'agri-businessmen' and speculators multiplied the economic incentives for intensive exploitation (Whyte 1998). By contrast, Spanish agriculture remained relatively unchanged until the 1950s (Simpson 1995). This recent modernisation may be a crucial element for explaining relative Spanish apathy with regard to GM crop plantings. Overall, however, the break-up of common lands for the benefit of the enterprising upper classes, exemplified by the British enclosures, was a broader European phenomenon that allows for some generalisations:

> The operating causes were everywhere the same: the trend towards the reintegration of large estates; the increasing emphasis on production as a private undertaking, with an eye always on the market; and the crisis among the rural proletariat, painfully adjusting itself to an economic system based on money and exchange.
>
> (Bloch 1966: 189)

Nowhere were the disenchanting effects of modernisation and its cultural riposte more tangible than in France, with its deeply ambivalent population – both modern and intensely traditional (Bess 2003). By the 1860s, technology and economic forces began to be perceived as threats to the traditional landscape (House 2000). French preoccupations centred less directly on aesthetics than on rural depopulation and spectacular technological changes after the Second World War:

> Patterns of life that had endured for centuries, even millennia, suddenly mutated or disappeared, like a movie in fast-forward. Populations vanished from the countryside; modes of farming changed and changed again; trade balances shifted; ancient villages became ghost towns; suburbs invaded the abandoned fields; high-tension wires and high-speed trains sliced through quiet valleys; the very face of the land was recast.
>
> (Bess 2003: 39)

Although they recognised its benefits, the great majority of the French people did not genuinely welcome the 'march of progress'. Farmers continued to regard themselves as 'peasants', as guardians of the countryside and 'rural artisans' rather than mere entrepreneurs, displaying an 'obstinate attachment to their old habits' and contributing to a slow, wavering pace of modernisation (Bloch 1966: 241). Specialisation and mechanisation were frequently seen as inimical to an ancestral, essentially 'multifunctional' way of life that included cottage gardens and poultry yards, stables and pig-sties. The lifeworld of French peasants was characterised by a mélange of economic and cultural rationalities, their work was both vocation and social identity, their connection to the land both affective and practical.

The peasantry's hold was not broken until the post-war phase of modernisation. A potent mixture of geo-political strategy and 'technological enthusiasm' menaced the old agrarian way of life (Bess 2003: 20ff.). The quest for national security and grandeur was widely supported, but the reliance on techno-science mirrored the contemporary case for agbiotech. Promethean industrial development ignored the identity-based link between present and past, between people and the land, between society and nature. Most French people felt and still feel a deep connection to the agri-cultural world and even sophisticated Parisians will flaunt their provincial roots. Many own a small country house, often inherited from rural grandparents, or regularly visit the diverse regions on 'agro-tourist' trips (Heller 2006). For decades,

> in memoirs, films, novels, newspaper articles, academic conferences, and countless town meetings, the French made it clear that they perceived the disappearance of the old rural world not just as an economic or demographic challenge, but as a spiritual loss, a deep wound in the tissue of their civilization.
>
> (Bess 2003: 40)

The 'double process' of intensification was in full swing here once again. Whereas French agriculture, buoyed by European subsidies, was turning into an industrialised export sector of the national economy, traditional virtues, the French *paysage*, and rural *douceur de vivre*, began to play an ever greater role in defining national identity (ibid.: 49).

Besides powerful signs of an 'agri-cultural' French identity, cultural ideas of nature also play an important part in the rejection of modernisation and, by corollary, of GMOs. Intimately linked to agriculture, what Heller (2004: 88–89) calls 'French nature' is 'predicated on romantic notions of *social* rural life'. It is a 'worked nature [...] whose value is not only historical, cultural or aesthetic, but also *economic*, providing *paysans* with a viable and productive way of life'. The slightly 'wilder' images of England's Lake District or German forests had never found many supporters among French Romantics who typically sought a '*marriage* of nature and culture' (Charlton 1984: 214). Most French citizens had an even more 'artisanal' and cultivated conception of the nation's nature. The vast stretches of fallow land resulting from rural depopulation were often seen as the most troubling signs of modern change. Even a leading green politician, Antoine Waechter, wrote in the early 1990s that fallow land was a sad image of 'grievously degraded territory' (Bess 2003: 134), demonstrating that 'French nature' stands in direct 'historical continuity with culture' and is embodied in an 'agri-cultural' way of life (Heller 2006: 330). Whereas 'wilderness' is invariably associated with nature in far-away countries such as South Africa, usual features of ecological 'nature', such as mountains or swamps, are widely recognised. However, the crucial insight of a 1991 survey on perceptions of 'nature', from which this data is taken, is the number of answers pointing to relatively tame elements: wheat field (88%); vineyard (86%); herb garden (84%); urban park (63%); tree-lined avenue (56%) (Bess 2003: 133–134). Many hybrid environments, such as orchards or fields, were actually preferred to 'a natural site untouched by man.' For most French people, then, going 'back to nature' signifies 'returning to one's rural or agricultural roots', to the mythical *terroir* (Heller 2006: 331).

When one of the foremost anti-GMO organisations – the *Confédération Paysanne* (CP), a powerful union of smaller 'peasant' farmers – invokes the protection of nature, it mainly refers to 'that which is not urban'. In the same way, French cheese like Roquefort is seen as 'natural-as-agricultural' in comparison to industrially produced foods, including GMOs (Heller 2007: 604).[2] Similar dynamics can be observed in Italy and other Mediterranean countries such as Greece, Cyprus and Malta. In Italy, the small farmers' organisation *Coldiretti* was at the helm of an alliance of consumer and environmental groups and sought to remind Italians that agricultural industrialisation represented an assault on their national agro-food patrimony (Kurzer and Cooper 2007). Organic agriculture has grown more convincingly here (than in France) by stressing its connection to high quality, traditional land management and good taste. In general, anti-GMO farmers, their allies and numerous sympathisers do not need to draw on images of nature as consistently as campaigners in northern Europe because the appeal of the 'natural' is inherent in agricultural stewardship and associated landscapes. Moreover, nature preservation does not necessarily stand in opposition to cultivation.

Agricultural values in America: Productivity or virtue

From today's vantage point, it may seem as if US agriculture never carried a comparable 'burden' of meaning and cultural identity, but that would be an oversimplification. While it could not aspire to the same mystical qualities, agriculture occupied a central position in the socio-cultural fabric until at least the early twentieth century. Many commentators have recounted the story of how the Jeffersonian ideal of the virtuous, small-holding and fiercely independent yeoman farmer appeared to flourish with the expanding frontier and was then smothered by an alliance of industrialist and progressive forces, united in a utilitarian embrace of technology and productivity (Bradbury and Temperley 1989; Thompson 2001). In the nineteenth century, images of well-ordered and productive gardening were still used to describe the Jeffersonian landscape and encourage its expansion into the uncharted West (Arnold 1996). Thus, there certainly was a degree of aesthetic or identity-related connection between farmers and their land, but the American virtues it embodied were linked to 'progress', hard work, and enterprising spirit – reasonably similar to the philosophy of industrialism. Unlike in Europe, agriculture did not usually stand for a paternalistic 'moral economy' and customary communal ties or embodied the ideas of natural order. Instead, it was part of a fundamentally antagonistic 'man-versus-nature' orientation that had no serious cultural leverage against the onset of large-scale mechanisation. The twentieth-century political movement of populism, chiefly supported by small farmers and landowners, only expressed a conceptually limited opposition to socio-economic change with the anti-corporate principle of 'big is bad' (Thompson 2001: 223). While British or French public life is still suffused with countryside values, traces of Jeffersonian agrarianism have been preserved in the new American social formation of 'suburbia'. The burgeoning suburbs, writes Tuan (1974: 237), are definite 'heirs to Jeffersonian family farm values.' They reflect a comparable kind of individualism and helpful neighbourliness on the one hand, and continue the theme of human domination over nature, symbolised by impeccable gardens, neat lawns, and pets in place of farm animals. The names of suburban, frequently 'gated' communities – such as *Pheasant Acres* or *Shady Grove* – 'evoke the rural ideal and nostalgia.' Rural virtue now resides in suburbia, while the actual countryside edges ever more into the world of agri-business. Reformers have not lost all hope for urban renewal (Kunstler 1998), but their successes do not yet amount to a powerful trend.

The work of Wendell Berry and a small but vocal counter-culture continue to promote an alternative, virtue-based version of agriculture founded on the principles of place-based identity, solidarity, and sustainability (Flammang 2009). Berry (1977: 26) proposes a more integrated way of life reminiscent of European traditions:

> To live, to survive on the earth, to care for the soil, and to worship, all are bound at the root to the idea of a cycle. It is only by understanding the cultural complexity and largeness of the concept of agriculture that we can see the threatening diminishments implied by the term 'agribusiness'.

Yet, Thompson (2001: 227–228) concludes that Americans are unlikely to adopt this recipe for a new ethical agrarianism because the utilitarian critique of health and environmental impacts of industrial agriculture remains the predominant theme. And this can probably be served by an ecological modernisation that reduces such undesirable externalities without changing the underlying structure. As Echols (1998: 529) observes, 'change and experimentation' have long been admired in the US, particularly when performed through new technologies and aimed at increasing food safety or raising productivity and competitiveness. A scientific and utilitarian outlook is promoted by USDA through its '4-H Club' youth organisation which, since the early twentieth century, had 'set out to plant science and capitalism in its farm kids' (The Economist 2013). The programme's summer camps and activities are designed to equip future generations of US farmers with both the knowledge and the values needed for running competitive 'agri-businesses' in a globalised world.

Finally, a structural explanation for the utilitarian perspective on agriculture is also found in most citizens' separation from the middle landscape of food production. Important exceptions certainly exist. In New England, for example, in addition to the presence of numerous organic farms, almost 'European-style' landscapes and farming systems have fostered an unusually strong 'cultural identification' with food and agriculture (Tokar 2009). In much of the US, however, the links between the rural and the (sub)-urban are tenuous, and there tends to be a sharp demarcation of farmland and wilderness. In a rural universe that is largely animated by economic competitiveness, the cultural politics of nature claims a permanent backseat (Herrick 2005).

Europeans and 'Environmental Identities'

The fundamental point illustrated by these various 'cultures of nature' and agrarian traditions is that most Europeans' local, regional and national identities are 'environmental identities' in the sense that they include their cultivated but partly 'natural' surroundings. Although many aspects of American national and regional identities are evidently 'environmental' as well, for they encompass 'natural' elements beyond the self, the hybrid 'middle' landscapes of work and dwelling are largely bereft of strong spiritual or identity-related attachment. By contrast, the French case stands out as one of the clearest expressions of what I labelled as the interactive 'ontological condition' of nature and humanity in Europe, whereby what counts as nature or culture becomes almost indistinguishable. As a consequence, what happens to surrounding environments has a direct impact on people's sense of identity.[3] The long historical struggle between modern technological transformations and a dogged attachment to customs, wholesomeness, authenticity, and to notions of nature is entering another round with the arrival of GMOs. As the first image of modernity 'colonises' more and more areas of life, the identity-related reaction becomes stronger and broadens its appeal, in some cases assembling significant popular majorities that cut across political boundaries. For many southern Europeans, the GMO revolution fits seamlessly into a succession of technological innovations, this time turning nature itself into a part of the capitalist machinery. In the 2001 Eurobarometer survey, France stood out as the most 'technosceptic' nation when it came to science and technology in farming. French campaigners successfully linked the concern for the agricultural 'soul' of the nation to the destructive effects of globalisation and economic liberalisation (Bonny 2003). The role of 'nature' here was clearly distinctive, as it usually referred to a fusion of culture and nature and relied heavily on the sense of a virtuous past which I termed 'nature as nostalgia'. Having emerged from dictatorial regimes quite recently, these associations obtained to a lesser degree in countries such as Spain or Portugal. There were some parallels with the German experience, as the idea of nature had been tainted through its profligate use by the ruling totalitarian elites (Sanz-Lafuente 2003).

What most southern Europeans share with their northern neighbours is the perception that identities and the lifeworld are under increasing pressure from a cascade of socio-cultural and economic changes. In Marris et al.'s (2001: 66) focus groups, respondents across Europe made it clear that many modern agricultural methods were considered 'unnatural', including pesticides, animal-derived feed and routine use of hormones or antibiotics. It makes little sense to interpret these attitudes as evidence of scientific ignorance or general hostility towards technology. Although the critique targeted the modern rationalist trajectory as a whole, it focused on a specific variety of nature–culture assemblages, and most prominently on biotechnologies. Most other technologies, such as computers or mobile phones, are rarely mentioned as a source of deep anxiety. It should also be noted that there is a greater reliance on explicit ideas of nature in northern European regions. I have already discussed how one of the most valued features of the British countryside is its wildlife, especially farmland birds. Ecological concerns thus play a more significant role in public discourse, yet they rarely aspire to the kind of 'scientific neutrality' projected by expert committees. Bird protection movements in particular, whose sympathisers number tens of millions across northern Europe, have historically accepted a cultural and cultivated landscape, while monitoring the ecological effects on bird populations (Schmoll 2005). An extensively managed agricultural system is frequently seen as the preferred long-term solution (Eccleston 2007).

Generally, the public's vernacular thinking goes well beyond ecological precepts and draws on values and identities. Environmental NGOs like Greenpeace appealed to perceptions of cosmological disturbance and added the powerful image of pollution, involving a hybrid of environmental/material and moral effects. They described GMOs as self-replicating pollutants, as 'genes out of place, an ominous "reconstruction of nature"' (Levidow and Carr 1997). Aquino and Hallman (2004) identified the main themes of German respondents as revolving around the 'ecological balance of nature' and the 'danger in constant change and the human influence on our food'. Compared to Americans' focus on specific physical risks (highlighted by the same study), this translated into much broader concerns which evoked the concepts of 'natural order' and human modesty. Despite a stronger emphasis on the negative externalities of agriculture, the valued landscape in this case was still 'agri-cultural' rather than 'wild', underscoring the underlying civilisational commonalities with southern European attitudes. Furthermore, the 'agri-cultural' appeal of the humanised, cultural landscape is not the only dynamic correlated with anti-GMO sentiment. In Italy, for example, the public took note of GM agriculture once 'Dolly the sheep' and cloning had captured the headlines (Mini 2005). A strongly moral, often religious stance was transposed onto GM crops and foods. A similar, albeit less well documented, trajectory unfolded in Greece, where public apathy turned into enduring moral outrage. There is thus little hope of fully disentangling the processes of spiritual and sociocultural resistance. The sole certainty relates to their identity-affirming properties.

Food culture(s): Identity affirmation and re-enchantment

The indispensability of food for human existence makes it a primary reference point for all societies across the globe (Van Waarden 2006). In this section, I examine the ways in which food transcends its nutritional role. It nowadays represents a major flashpoint in the process of modernisation, injecting its symbolic potential into struggles over identities, visions of the good life and, ultimately, the biotechnology revolution.

Food and identity: Bridging the nature–culture divide

Historically, the foods consumed served as a mirror of a society's environmental circumstances, of the manner through which nature was converted into edible, life-giving substances. The menu of traditional societies closely reflected their seasonal produce and access to animal protein, but it also went beyond such pragmatic aspects. Just as medieval peasants saw their natural environment as a utilitarian-spiritual hybrid, foods have long been subject to both religious principles and considerations of safety or nutrition. Stanziani (2008) cites the Jewish injunctions of kosher food preparation – related to safety aspects but articulated in religious terms of purity – to illustrate the compatibility of religious symbolism and cautious food innovations. Since ancient times, kosher meat was a way of ensuring the freshness of meat products. These days, rabbis need to consult scientific knowledge to judge on the admissibility of modern colouring and preservatives. There is much at stake in the act of eating, as this most basic, daily practice is at heart 'a rather risky business', both physically and culturally (Fitzgerald and Campbell 2001). The problem of food adulteration has been a foremost concern ever since a rudimentary barter economy allowed for trade and specialisation (Van Waarden 2006). The relative proximity of food production, buyer-seller relations marked by trust, and supervision by medieval guilds mostly kept this murky business in check. Even before the nineteenth century, the connotation of 'naturalness' was frequently associated with the trustworthy and safe provenance of food items (Stanziani 2008).

Besides the purity-safety linkage, food also constituted a focal point of the moral economy which I described in Chapter 5 as the socio-economic complement to the spiritual 'natural order'. The gradual shift towards a political economy based on free market principles deprived many people of an underlying safety net provided by Church and state authorities. In times of crisis, the Catholic Church often tried to force merchants to lower their prices to an acceptable level, while societal elites were expected to live up to their 'paternalistic responsibilities' and provide food for the poor (Pilcher 2006: 43). Over time, customary rights were stifled by market-friendly laws. As national and international markets increasingly gained importance, the key constraints on food and cuisine shifted from the natural environment (soil, seasons, weather) to the logic of commodity trading where mainly price and quality (or shelf live) counted. The mass production of foods, many of them canned or frozen, which was pioneered by American corporations, brought about a growing market of largely 'interchangeable commodities' that had lost 'all connection to their place of origin' (ibid.: 55).

The process of culinary industrialisation undermined public trust in food safety and swept away the vestiges of a moral economy that treated food according to a framework of entitlement and obligation. From the 1870s onwards, a proliferation of state agencies, trade associations and legal statutes, such as the British Adulteration Acts, instituted regular controls to foster greater public confidence. A burgeoning advertisement industry tried to entice consumers to switch from traditional produce to 'new and improved' modern food items – nowhere more successfully than in the US (Levenstein 1988: 43). Yet, none of these developments took shape without significant controversies and public mobilisation. Little else in human societies has the same capacity as food to arouse discontent and distrust. The tightening of government regulations by the end of the nineteenth century was only the most visible consequence of a primarily middle-class campaign against the corporate take-over of food cultures. They may have been 'heirs to the moral economy of eighteenth-century food rioters' (Pilcher 2006: 59), but they also gave the defence of food a Romantic twist, setting it up as a symbol of traditional authenticity in the context of growing market relations. Cultural and political mobilisation around food always pays attention to the dual nature of food, both material sustenance and cultural signifier. The 'bio-cultural' character of food politics and nutritional science (Fieldhouse 1986) is broadly recognised in the literature. The significance of food products and dishes has historically not been unavoidably tied to their bio-chemical make-up and life-giving properties. Rather than reflecting mere survival instincts, food is always deeply influenced by human practices of 'meaning-making'.

In the post-scarcity context of contemporary Europe and America, one should expect the cultural or symbolic role of food to be considerably magnified. As Fieldhouse (1986: 44) concludes, 'where there is more than sufficient food available for survival, choices are made, which are assertions of identity.' In the early 1960s, late modern societies were beginning to pay less attention to material and productionist aspects of food, although this was not immediately reflected in public policy. Of course, the same does not apply to most other parts of the world where food and agriculture remain closely linked to environmental justice, socio-economic development, and physical survival. Nor is post-productionism a prevalent attitude found among farmers themselves. In the industrialised world, at any rate, the abundance of food items made discriminatory choices and associated signs of distinctiveness more attainable than ever before.

Food cultures as a heuristic framework

To conceptualise the domain of food, culture and identity, I employ a 'possibilist' framework, similar to the one I proposed for cultural politics (Chapter 3). Human behaviour is shaped by socially constructed habits which, in turn, are never fully determined by material circumstances. The notion of 'food culture' can deliver such an approach, referring to 'a constellation of socially produced values, attitudes, relationships, tastes, cuisines and practices exhibited through food' (Lang and Heasman 2004: 185). Of course, food cultures can also account for intra-societal diversity, such as class-based ideals of fancy dishes or desirable body images linked to food consumption. In other words, food cultures cover both the 'social cement' and 'opportunities for difference and distinction', while still indicating an underlying commonality (ibid.). Regional or national food cultures are never monolithic constructs and their evolution over the past centuries teems with movements and counter-movements, fashions and experiments. Overall, however, it makes sense to locate food cultures as an essential component of Eder's 'double process' of intensification – the industrialisation and mass production of food as opposed to the veneration of 'natural' products and the return to traditional culinary roots (Eder 1996; Murdoch and Miele 1999). Nutritional science has long sought to guide food choices, but this has not overly constrained existing food cultures. The British infatuation with the 'Mediterranean diet', for instance, is not purely based on potential health benefits. It equally reflects the longing for the dream of sunshine, beach and the easy (yet cultured) life on southern Europe's shores. The modernisation of food through standardisation, harmonisation, globalisation and efficient manufacture – the wholesale commodification of food – is the process which most heavily clashes with the identity-related associations of food. By reducing or even eliminating its propensity to carry connotations of social distinction, religious symbolism, social belonging and cultural identity, the economic-utilitarian perspective sets in motion another wave of 'disenchantment'.

The perception of food as a tradable commodity is integral to an economic perspective and embedded in regulatory market institutions around the world. It has long been part of the WTO's mission to 'normalise' the status of food in the global trade system and reduce the political clout of 'non-trade concerns', ranging from food security to agro-biological diversity (Johansen 2007). A consolidated global economic order is to be built on objective scientific risk assessment and leave the exercise of choice to individual consumers. One can further observe this first image of modern food cultures in the growth of an oligopoly of food retailers and processors as well as the steady advance, until recently, of industrial farming, both spurred by economies of scale (Jacobsen 2004). From a cultural perspective, the lifeworld is now more frequently reduced to a one-dimensional logic. Food comes to be regarded as 'a matter of efficiency and an expression of technological know-how. [...] People see food as fuel, to be ingested with dispatch so as to make time for something else' (Visser 1999: 123). 'Food as fuel' represents not only the rationalist colonisation of cultural practice, but also spawns its own ultra-utilitarian cultural variant of 'fast food'. America's iconic culinary invention, the burger, has become a 'symbol of modernity and the triumph of an American mode of eating, a metaphor for how a rational, bureaucratic society can command the production of any good and the fulfilment of any need' (Lang and Heasman 2004: 191). From an extended historical perspective, 'food as fuel' is also heir to the Enlightenment vision of progress and an expanding sphere of human control. With its promises of 'solving' global hunger and liberating humanity from the daily battle against nature's armies of insects and weeds, GMOs equally represent a continuation of the Enlightenment project. Nutrition and abundance are its foremost arguments, while improved taste, a tentative step into the non-utilitarian realm, has so far failed to enthuse most consumers, particularly in Europe. In the US, the creative and problem-solving possibilities offered by engineered foods have been received more positively (Kniazeva 2005).

Three variants of 'food as nature'

The first image of modern production and consumption with its trajectory of industrialisation creates significant anxieties, partly because it is seen as leading to recurrent disasters such as mad cow disease. Reconnecting this scepticism with the trump card of 'unnaturalness', Macnaghten and Urry (1998: 264) note that modern agro-food practices 'are increasingly viewed as immoral, wrong and "unnatural".' This flight to 'nature' has been a long-standing trend in the industrialised world (including non-Western countries like Japan or South Korea) and offers a trusted advertisement strategy. In the US, the 'natural' movement formed part of the counter-cultural 1960s and was quickly assimilated into product marketing, whereas in Europe nature had already become a 'byword for authenticity' in the 1920s, with companies capitalising on the 'association between nature, purity, simplicity and goodness' (Coates 1998: 9).

As in the earlier analyses of landscapes and agriculture, the term 'nature' designates a panoply of attitudes to food. Spiritual tendencies are more frequently found in northern Europe where nature as an autonomous spiritual force has joined a heightened interest in the ecological complexity of the natural world. The cultural context here emphasises the scarcity of nature as compared to the relative abundance of food. Nature is increasingly valued for its own sake and 'naturalness' is perceived as a 'quality criterion' without automatic reference to taste or other specific functions of food (Peters et al. 2007: 198). This discursive field is cultivated by environmental NGOs which 'paint a picture of a nature tortured by human ignorance, exploitation and misuse, a gloomy picture of paradise lost. Poisonous foods and famines are part of the price we pay for our scientific and technological hubris' (Jacobsen 2004: 63). The clash between scientific-industrial society and a set of cautious, wholesome, 'common-sensical' and natural methods manifests itself in organic foods and other 'naturally' branded products, on the one hand, and conventional 'productivist' food products on the other. While public health and the fear of 'nature as boomerang' (Murdoch and Miele 1999) are always part of this approach, its rejection of industrial methods and GMOs, in particular, goes well beyond an emphasis on physical risks.

Associations of spiritual purity and moral goodness are partially subsumed into the attitudes that predominate in southern European countries. Yet, here the 'humanised' nature of artisanal production methods and traditional recipes constitutes the most prevalent understanding of 'natural' foods. Jacobsen (2004: 71) writes that 'food is profiled by reference to farmers' traditional and "natural" practices, wildlife as well as hunters' primeval instincts, the butcher's traditional professionalism and mother's cookery (based on grandmother's recipes).' The standardisation of taste and the loss of varied and fresh produce are further complaints which are often articulated in the south (Zechendorf 1998). Marked by a classical agrarian outlook, southern European societies tend to appreciate the peasants' vision of nature as an 'embodied experience' and savour food as nature 'improved' through revered artisanal techniques (Jacobsen 2004). In this symbiotic worldview, producers and consumers largely adopt a common standpoint, while ideas about abstract, spiritualised nature have less resonance than in northern Europe, unless they connote a religiously legitimised 'natural order'. As I observed earlier, the perceived violation of this 'natural order' was the most immediate trigger of public outrage in many southern European countries.

A third role of nature in European food cultures is most widely shared and effortlessly bridges the north–south divide. Nature as a signifier of a golden past applies to food in a particularly acute way. Unnaturalness here functions as a linguistic placeholder for seemingly uncontrollable socio-cultural changes. The decline of home cooking and the family meal, the capitulation of taste before shelf live, the multiplication of 'new and improved' or 'quick and easy' labels signal, at the level of everyday life, the advance of economic efficiency and specialisation at the expense of 'glacial' time (Macnaghten and Urry 1998), and the pleasures of a simpler, more wholesome life. In focus groups across Europe, modernity was recognised as bringing some benefits, but this did not preclude a frequently made, negatively tainted distinction between 'traditional and modern lifestyle orientations'. The discourse of 'lost times' was used by both 'natural' and 'convenience' eaters, and across all age groups (Marris et al. 2001: 68).

Food and identity: The communal dimension

The multiple roles of food in modern societies evoke the earlier analysis of agriculture, which explains why most scholars treat agro-food questions as a single issue-area. This avoids duplication, but risks overlooking the distinctive aspects of each subject. In the case of food, heightened concerns about safety and nutrition are joined by an especially strong cultural symbolism. Agricultural systems could potentially be ignored by city dwellers in the industrialised world, but food consumption forcibly reconnects them to the cultivated landscape, prevalent farming methods and attitudes towards the natural world. I continue this line of argument by linking taste and quality with cultural identity before tackling the food-nation linkage and elaborating its place in popular demands for an alternative modernity.

Taste as a function of quality and culture

Modern food innovations, from frozen ready meals to packed sandwiches, not only advertise their convenience, but also emphasise an almost 'scientific' concern for great taste. With regard to GM food, Europeans have so far been unimpressed by similar promises. The pretensions of better taste (and nutrition) are not easily integrated into existing European food cultures where taste is closely associated with provenance, cultural history, and family traditions. Taste is as much socially learnt as it is symbolically charged. Wagner et al. (2001: 89) call better taste a 'non-issue' because it implies a departure from the existing menu of national or regional dishes. So deep is the cultural attachment to distinctive products and dishes that Italians, for instance, profess an unusual trust in their politicians' handling of food-related issues (ibid.). The visceral, shared bonds regarding Italian food trump an otherwise ingrained distrust. A similar relativity can be observed with regard to food safety, although again more in the south where raw milk cheese and cured meats are much appreciated. As microbiologist Richard Lacey (quoted in Bishop 1991: 32) comments, 'the French actually regard food poisoning as an acceptable risk.' By contrast, the same products are banned by US legislation (Echols 1998).

As these examples show, taste and cultural identity (and even risk perception) are very closely connected in some European countries. This linkage is certainly weaker in the European north, but even here one finds popular reverence for trademark dishes and beverages. Sassatelli and Scott (Sassatelli and Scott 2001: 224) opine that the British debate 'has not included much reference to traditional, local and artisan production and much less attention is paid to homogenization of flavours and loss of cultural identity than to animal welfare.' However, at least since the 1990s, northern Europeans have begun to re-acquire the kind of 'food values' still widespread in the early twentieth century or which they would observe during their regular holidays in the south. Celebrations of local and regional foods, for instance through food markets and festivals, are now commonplace (Jacobsen 2004). While northerners still retain a strong emphasis on utilitarian aspects of safety, nutrition and price, the revival of a more authentic culinary culture boasts a growing following among the middle classes. To varying degrees, culinary specialties 'constitute a part of the social fabric' (Gaskell et al. 2001: 111) in all European countries and this tendency has created considerable pressure for 'cultural protectionism'. That call has been answered by national and European legislations on labelling (Protected Geographical Indication) which reliably identify food products that come from particular regions and are made with traditional methods (Zurek 2007).

Food and the nation: A new sacredness

Historically, food is inevitably associated with the rise of the nation-state. The initial rejection by many Englishmen of potatoes ('Ireland's lazy roots') and 'fancy French dishes' or the Scots' patriotic decision in favour of oats (instead of 'English' wheat and barley) finds a contemporary expression in an uncomplimentary labelling of foreign nationals (Fischler 1988; Bishop 1991).[4] Some go as far as to suggest that food should be considered 'on a par with language in terms of cultural definition' (Bishop 1991: 32) because food consumption represents

> the basis of collective identity and, by the same token, of otherness. Food and cuisine are a [...] central component of the sense of collective belonging. In some situations of migration or of minority culture, it has been observed that certain features of cuisine are sometimes retained even when the original language of the culture has been forgotten.
>
> (Fischler 1988: 280)

Put differently, 'identification with a particular community is literally ingested' (Sassatelli and Scott 2001: 215) and renewed through daily practices of cooking and eating, frequently in a social setting which further reinforces the communal quality and cultural authenticity of food. Consequently, the identity-related properties of food have also been employed in nation-building strategies. One element of emerging national food cultures was the celebration of certain staple carbohydrates in art, religion and folktale. 'Unless they have eaten some of this basic food', conjectures Bishop (1991: 32), 'whether it be wheat, oats, barley, rice, potatoes, people traditionally do not feel that they have eaten.' Over time, with the greater availability of a variety of foodstuffs, this feeling may have weakened, but it is still difficult to imagine Italian cuisine without wheat pasta, British cuisine without potatoes or a French meal without the obligatory baguette.

A major dilemma of nation-building strategies was that food culture did not only unify populations, but also tended to split them into socio-economic classes and subcultures. Gastronomic writings during the nineteenth century therefore aimed at public education for quality cooking and sought to boost 'a sense of national identity and superiority' (Sassatelli and Scott 2001: 215). Yet, the tension between nationalisation and social distinction made the creation of national cuisines a difficult enterprise. It was accomplished in America by a move to mass-market industrial cuisine, but Europeans' efforts were long frustrated by the dual barriers of regional diversity and elitist aspiration (Pilcher 2006: 52–53). The French managed to find a compromise by accommodating an exclusive dining culture alongside a broader national food culture. A new category of national heritage, the *patrimoine culinaire,* was born and the French configuration was widely emulated across Europe (Knowlton 2007). However, the culinary diversity within national food cultures could never be held at bay and culinary change always took place – the reluctant introduction of potatoes is a historical example, the widespread consumption of Asian rice varieties and fresh exotic fruits a more contemporary illustration. Over time, especially in post-war Europe, features such as provenance and methods of cultivation and processing have become universal reference points for quality food: its appreciative adjectives range from natural to traditional, from regional to artisanal, from hand-made to home-made. Heavily processed foods may still constitute a national staple food, but a tin of baked beans is unlikely to be invoked as a proud embodiment of British national identity. Special beef dishes, on the other hand, would certainly qualify.

Food as counter-modern authenticity and stability

Despite the impression of cultural stability, there have thus been genuine changes and innovations in European food cultures. Yet, the more these changes seemed to emanate from external powers or corporate expansion, the more controversy they have generated, for reasons of both cultural authenticity and economic (dis-)advantage. Generally, the politics of food around the world is marked by a 'conservatism of food habits and the desire of people to maintain their traditional sources of sustenance' (Pilcher 2006: 119). So close is the connection between food cultures and identity that the introduction or invention of novel foods is bound to have a cultural impact, as Visser (1999) illustrates with the heavily advertised popularisation of cornflakes and milk as a 'normal' breakfast in what was once a milk-scarce Spanish food culture. Her examples of more harmonious and voluntary adoption are equally telling: tomatoes soon found a place in Italian food culture as an ideal complement to pasta dishes, while Americans more recently developed a fondness for chilli peppers, providing flavour 'easily, quickly, and indisputably' (ibid.: 121). In the culinary, cultural context of European nations, GM food resembles more the Spanish example of corporate promotion and socio-cultural destabilisation.

Gunn and Tudhope's anthropological study of biotechnology's impact on indigenous peoples contains an important caveat. In their richly symbolic and often animist universe, the introduction of GM foods 'risks creating cultural change, especially in cultures whose primary values are non-material' (Gunn and Tudhope 2006: 3). This message finds a limited echo in the European cultural politics of food and agriculture. Although Europeans' 'environmental identities' and culinary symbolism are not directly comparable to non-Western cultures, in the previous chapter I argued that medieval currents of thought still influence European reactions to the modern age and that settler nations like America are perhaps the strongest 'outliers' in world history. 'Food and modernity have shared a complex, often troubled relationship', muses Pilcher (2006: 118), which, however, neglects the possibility that food might become an expression of the modern *Zeitgeist*, as happened to some extent in the US.

The widespread unease in Europe about the industrialisation of core repositories of meaning, such as cultural landscapes, also applies to the production, preparation and consumption of food.[5] The utilitarian benefits of affordability and food safety have, to some degree, shielded the food sector in northern Europe from these concerns, but even here considerations of taste and authenticity are on the rise again, whereas safety has become doubtful. In Britain, a whole string of recent agro-food scandals, ranging from BSE to foot-and-mouth disease and beef mixed with horse meat, has typically been blamed on poor controls, commercial greed and the heavy industrialisation of farming (Sassatelli and Scott 2001). The originally positive connotation of 'cheap food' has increasingly come under attack and the supermarkets' stranglehold over their suppliers has been linked to everything from bland taste to food contamination and even to the obesity crisis (Nerlich 2004). 'They have high-quality, dangerous food', Richard Lacey remarks in his investigation of French cuisine. 'Other countries have medium-quality, safe food. We [British] have the worst of both worlds, bad quality, dangerous food' (quoted in Bishop 1991: 32). Not just the European south is reacting to the perceived homogenisation of taste, excessive commodification and overproduction of low-quality food. Italy's famous 'Slow Food' movement is only the best-organised expression of a pan-European trend towards an alternative culinary culture that rekindles traditional practices as much as it creates new ones, such as vegetarianism or veganism (Murdoch and Miele 1999).

> [A]fter decades of fast food, there is a fast-growing 'slow food' movement; after decades of legal adulteration of food, there is now a burgeoning market for natural foods; after decades of enticing

> consumers to eat world cuisines, there is now a counter-move to return to localism, regional foods and real cooking.
>
> (Lang and Heasman 2004: 189)

In southern Europe, despite a rising popularity of fast food, the positive connotations of old-style, 'genuine' food are well-ingrained and often associated with celebrated national cuisines and regional peasant traditions. Food scandals are generally low-key and swiftly forgotten, unless they pertain to industrial food production. The fusion of food and national identity has attained such a degree in Italy that unwanted innovations (including GM food) are regarded as 'alien' and therefore 'un-Italian', besides presenting a threat to artisanal production methods and the reputation of Italian products (Sassatelli and Scott 2001: 227).

Food as resistance to globalised placelessness

Cheap, mass-produced food, and GMOs in particular, are identified with another all-pervasive transformation that causes much anxiety. While many Europeans have become used to the idea of importing low-cost electronic goods from Asia, tolerance of globalisation often hits the buffers when core aspects of national cultural identity are at stake. For a long time, national food regulations protected the culinary heritage, but this stability is now in doubt. European harmonisation of food standards and safety requirements as well as international regulatory institutions, such as the WTO, have enfeebled national autonomy and re-regulated markets, often in the interest of large agricultural producers and multinational food corporations. Even the European Consumers' Organisation (BEUC), which should in principle favour ever wider 'consumer choice', agrees with placing some cultural limits on the global trading system (Zurek 2007: 364). Moreover, the spatial dissociation of consumers from the origin of food products and uncertainty about production methods enhance the commodity status of food while undermining its traditional signification: 'Modern food has become [...] an "unidentified edible object", devoid of origin or history, with no respectable past – in short, without identity' (Fischler 1988: 289). Globalisation 'breaks up the congruence between the territory of societies, markets, cultures, and regulatory states' (Van Waarden 2006: 50).

Worries about the disappearance or dilution of food cultures are expressed through opposition to 'alien' or 'unnatural' foods, on the one hand, and through critiques of US 'cultural imperialism' or 'burgerisation' which are articulated by actors within the anti-globalisation movement (Knowlton 2007). The French term *la bouffe* (literally 'grub' in English) encapsulates the celebration of national dishes and artisanal products in the face of global homogenisation. *La bouffe*, explains Heller (2004: 92), brings together 'notions of pleasure, tradition, and French cuisine, synonymous with French culture itself.' The 'peasant' coalition, led by the José Bové, which spearheaded the French anti-GMO movement, drew heavily on the antithesis as well (Gordon and de Boisgrollier 2005). *La malbouffe* designates 'everything distasteful about globalization ranging from the cultural homogenization associated with McDonald's fast food to the industrialized agriculture associated with hormone-treated beef' (Heller 2004: 92). Such feelings are amplified in much of southern Europe, though somewhat muted in newly modernising societies, such as Spain and Portugal, where a return to tradition evokes uneasy memories. But the identity-based embrace of food cultures also exists in the north. Certainly less ferocious than southern resistance, the widely supported Campaign for Real Ale, the rediscovery of regional cheese varieties, and a proliferation of products with protected geographical indication in the UK nonetheless share the pursuit of authentic quality and cultural symbolism, while challenging an economic logic of 'artificial' branding and globalism.

Leitch (2003) suggests that this type of protest has grown substantially since the early 1990s, constituting a new form of political engagement. Globalisation, increasingly conflated with European integration itself by European citizens, is coming under the spotlight. A return to the 'roots' is often proposed as a way of resisting the global juggernaut of homogenisation. As noted by Sassatelli and Scott (2001: 239), tradition is more and more touted as a 'solution to "modern" problems', for instance by active state support for national/regional food cultures (and consumption patterns) that allow traditional landscapes and their resident food producers to survive, even in more liberalised markets. Admittedly, not all European countries are equally engaged in similar agro-food strategies. For every front-runner (such as Austria), there are also regulatory laggards like Britain, which favours a hands-off approach, or the Netherlands with a history of highly technological modern farming and export earnings premised on low price and safe produce. Overall, despite a relative north-south division, the communal values and identity-related signatures embedded in food cultures play an important role across Europe. Just as agriculture is not merely another sector of the economy, food and cuisine are far removed from being judged solely on utilitarian grounds of safety and affordability.

As a symbolic amplifier and guardian of national cultural identities, food cultures are unsurpassed.

> The 'natural' and 'grandma's' style of preparing food becomes a shorthand for identity, for sensing oneself as belonging to a cultural group, a feeling that goes far beyond any consideration of nutritional value, WTO trading rules, or practicalities in industrial food production.
>
> (Wagner et al. 2001: 89)

From this perspective, the trouble with GM food is that it does not merely constitute a new product line which might somehow find a niche in the market. Instead, it embodies a new process of production which alludes to the very opposite of what food cultures seek to preserve. Without a strong rationale for their adoption, GM foods continue to be associated with the realm of industrial food production. As Wagner et al. (ibid.) put it, 'GM technologies are seen as anti-cultural. At a time of an abundance of food in the West they are seen as unnecessary.' To culinary, cultural identities, they are a threat, not a promise.

America: A truly modern food culture?

The question is, of course, whether America's food culture can be distinguished to the same degree from Europe as in the case of 'cultures of nature' and agricultural systems. After all, the US may have gone through similar phases of customary food culture, modernisation and re-traditionalisation. While American food culture has some commonalities with northern Europe – especially regarding food safety, nutrition, and price – its modern materialism and the victory of convenience food differentiate it from the richly symbolic, traditionalist culinary worlds that still characterise most European societies. The oft-bemoaned state of affairs in the US is articulated by Knowlton (2007): most American citizens cite modern fast food like pizza, tacos or hamburgers as national dishes, applaud the spread of these foods around the globe, and regard serious cooking as a leisure activity. Food arguably occupies a rather banal place in society, if some shopping malls sell certain products by weight – without restricting such offers to pet food or fruits and vegetables (Toke 2004: 98). From a historical perspective, there certainly was a 'pre-industrial' time when food expressed traditionalist cultural meanings, when the guiding theme was not individual choice but communal identity and faithful custom. Until the late nineteenth century, a modified Anglo-Saxon food culture with an emphasis on opulent beef dishes was still very much in evidence. Native or immigrant cuisines were sidelined (even before American independence) and, contrary to today's metaphor of the 'melting pot', great strides were made to assimilate immigrants into the existing culinary universe, if often unsuccessfully:

> As migrations peaked in the first decade of the twentieth century, educators sought to indoctrinate the new arrivals into the local culture, including pot roast and baked beans. One social worker reported a visit to an Italian home with the words: 'Still eating spaghetti, not yet assimilated'.
>
> (Pilcher 2006: 69)

Reports from European travellers highlighted some perceived differences with European food cultures. Almost all visitors were shocked by the abundance of food on the tables, by the quantities consumed, and by the amount of left-over food going to waste. British writers observed that 'Americans tended to eat more corn, pork, molasses, and indeed [...] much more of everything than did the British' (Levenstein 1988: 3). Other defining characteristics included a heightened taste for sweetness (partly to counteract excessive saltiness) and a weakened social function of food, 'manifested in a tendency to eat and run, rather than to dine and savour', and in an 'eerie silence that reigned at American dinner tables' (ibid.: 8). Regional cuisines and elaborate dishes certainly existed in the New World, but a new narrative of industrial progress and of liberation from daily toils would gradually undermine both diversity and artistry, while remaining true to the self-understanding of the 'people of plenty'. The twentieth-century nationalisation of American cuisine did not raise the standards and glorify authentic ingredients like it did in France, for example. Instead, it ushered in the 'mass market of industrial cuisine' (Pilcher 2006: 53) where the preference for beef carried on in new forms such as the fast food burger.

The rise of the modern food industry in America underscores the close relationship between agriculture (production) and culinary culture (consumption). As the former became big business by the turn of the twentieth century, its expanded geographical reach (through the railways) and enhanced processing technology mirrored the transformation of other economic sectors. 'Standardized hamburger flipping and French frying', notes Pilcher (2006: 108), 'grew out of time and motion studies conducted in the early 1900s by industrial efficiency expert Frederick Taylor.' Technological prowess, however, presented its own problems. It needed mass consumption and growing markets for the new food industry to prosper. Hence the advertisement machinery sprung into action to stimulate people's needs and promote new social customs, winning over the American people with images of modern abundance and joyful consumerism. In a sense, modernity's cultural appeal seemed like an unstoppable force. To the majority of rural or small-town dwellers, 'the artificial-looking, lightly colored, packaged brand-name foods of the city connoted purity, prestige, science and being part of the wider, more sophisticated urban centered culture' (Levenstein 1988: 201). This trend towards highly processed 'convenience foods', as opposed to superior ingredients and taste, was reinforced by a growing post-war infatuation with individualistic lifestyles, focused on independence and leisure at the expense of food preparation and sharing in family settings (ibid.: 162).

Considering these historical impressions of American food culture, the contemporary examples presented at the beginning of this section appear to be a logical outcome. However, it may be exaggerated to describe mainstream food culture solely as a reflection of basic modernist themes such as efficiency, technological progress and (auto)-mobility (e.g. Krenzler and MacGregor 2000). These motives are complemented by symbolic associations and images of pleasure. As Fieldhouse (1986: 13) cautions, a clumsy 'materialist' advertisement such as 'Crispy crackers fill your stomach fuller than other products' was still unthinkable, even if it referred to nutritional facts. Some foods (e.g. low-fat, probiotic) may be marketed almost like medical drugs these days, but their implicit signals of health, vitality and beauty are themselves embedded in a framework of basic American cultural themes: capitalism, industrialism, democracy, pluralism, individualism, leisure, and youthfulness (Jerome 1977). The new social needs of American consumers may still be similar to what would be considered fundamentally *utilitarian* rationales (convenience, pleasant taste, price, health), as Visser (1999: 122) showed with her example of chilli spice: it fits in well with the above cultural themes – 'it is addictive, simple, obvious, cheap – different from everything else, but always the same.' Crucially, this insight about 'cultural fit' may be applied to product marketing as well. Rather than creating new needs in the first place, advertisers may further develop and tweak existing themes, not unlike my earlier distinction between underlying cultural values and creative discourses (Chapter 3).

One basic rule of modern American food culture, however, is the product of insistent marketing, mixed with a dose of curiosity and faith in modern progress: the ideology of 'new and improved' has proven its permanent relevance (Fieldhouse 1986). It is also highly relevant to the agbiotech controversy. If an industrial food culture has taken root to such an extent, and if the ideational-nostalgic power of agriculture is weak, what would lead Americans to reject GM foods and crops? The public debate has often focused on environmental and health concerns, but Kniazeva's (2005) qualitative study demonstrates that a theory of symbolic consumption offers a vital addition. Most consumers form an opinion by drawing on cultural and personal or emotional resources. The utilitarian instinct of progress and nutritional or environmental improvement is strongly represented in the discourses. Even those rejecting GMOs for being 'against nature' appear to be open, at least in principle, to 'better strands of seeds', if these can truly outperform 'natural' varieties (ibid.: 32–33). Others are so enamoured with the potential of 'mixing and matching' new genetic traits that they blissfully imagine novel solutions for their personal problems. In short, the great majority of Americans have a largely utilitarian or 'instrumental' attitude to food (Food Ethics Council 1999), even if expressed in personal, emotional terms.

Finally, this account of American food culture must also highlight a sizeable 'counter-cultural' movement which supports different variants of sustainable/organic agriculture as well as ethical, high-quality, and healthy alternative food cultures. The movement's roots mainly lie in the 'neo-romantic youth rebellion' of the 1960s 'which extolled all that was individual and natural and denigrated the mass-produced and artificial' (Levenstein 1988: 204). However, this cultural-political upheaval has not yet genuinely transformed American food culture, as the movement remains relatively 'ghettoised' (Toke 2004: 144). The food industry has quickly caught onto the new trends and co-opted its marketing potential, neutralising its radical edge in the process. There has certainly been a moderate resurgence of regional cuisine and food quality over the past few decades, especially on the East coast and in the south, often linked to celebrations of regional/local identities. The central point remains, however, that the overarching national identity accommodates, and partly even embodies, industrial food culture: critiques never refer to convenience food as being somehow 'un-American', in stark contrast to (especially southern) European food-nation linkages (Heller 2007).

The counter-cultural movement represents the bulk of anti-GMO activists in the US whose growing, but still limited influence I examined in Chapter 4. Potentially salient 'events', such as the spread of 'superweeds' (resistant to GM crop-specific herbicides) and

'GMO-contaminated' food supply chains, have so far failed to create sufficient popular moral outrage in most US states. The anti-GMO movement may eventually mobilise a critical mass of citizens in favour of 'consumer sovereignty' and mandatory labelling, and this may also lead to a more cautious scientific and regulatory approach to agbiotech. But it seems unlikely that the movement's central concerns about socio-economic and ethical consequences of GMOs will be taken on board. That would constitute a radical departure from a long tradition of innovation-friendly, market-based regulation founded upon the permissive principles of 'sound science' rather than explicit political and cultural values (Gaskell et al. 2002). A revamped US regulatory framework might in the end resemble a watered-down version of European legislation on agbiotech.

Conclusion

This final chapter built on the historical dynamics analysed in the previous chapter and has highlighted the relative similarity of past 'cultures of nature' and contemporary identity politics. Whereas more straightforward, utilitarian considerations may yet begin to play a stronger role in public attitudes to agbiotech, one should not ignore the profoundly cultural motivations that underpin Europeans' objections to the new technology. The symbolic referents of 'nature' and 'unnaturalness' are multivalent and have mutated over the centuries, but the ontological structure of interactivity with human society and basic civilisational dispositions have persisted, ensuring that identity politics is always 'environmental' in character, comprising landscapes with spiritual and/or socio-cultural meanings. Agriculture and food are arguably the two most prominent meeting grounds of nature and culture. Their cultural disturbance by the agbiotech revolution has provoked widespread rejection, not merely on a symbolic level, but also because the technology can easily be construed as a manifestation of undesirable developments, representing a 'sounding board' (Torgersen et al. 2002) for diffuse, yet enduring anxieties about modernity, technological quantum leaps, and globalisation. From a historical perspective, this type of concerns is not genuinely new. The disenchantment of the world has been an ongoing process for several centuries. It therefore seems sensible to situate the current controversy in the *longue durée* of struggles between lifeworld and system, between cool-headed rationalism and the claims of custom and belonging. Notwithstanding internal European differences, in the eyes of many Europeans, signatures of meaning and identity, reinforced by nationalist mythology and embedded in agriculture and food cultures, are seen as threatened by rapid modern change. In America, although change has never been automatic or uncontested, it has generally been embraced more willingly and even celebrated at times, as modernist visions of human progress and industrial innovation have colonised the middle landscape of agriculture and the culinary universe.

Conclusion

This book has made the case for embedding existing perspectives on US and EU agbiotech regulation in a broader cultural-political account that complements agent-centric analyses with contextual insights. By highlighting the catalytic potential of cultural-political opportunities, I have shed light on the deep-lying sources of regulatory divergence and, by extension, on the persistence of the transatlantic divide. I have argued that environmental history and macro-sociological accounts can elucidate contemporary responses to a particularly controversial environmental (and moral) topic: the genetic modification of 'nature', which is a project to 'redesign' and commercialise biological life (Tokar 2001). Existing perspectives on regulatory politics deliver multi-causal analyses which are essential for a comprehensive explanation. Political economy approaches thus identify public opinion, political mobilisation, short-term interests, and the structure of commodity chains as critical elements. Institutionalists draw attention to political opportunity structures (such as political systems), bureaucratic politics, and the gradual entrenchment of organising principles and regulatory traditions.

However, neither approach fully explains the trajectory of the US debate over agbiotech which has seen a new wave of activism since 2009. Nor do they adequately account for why Europeans' concerns over agbiotech were tangible well before the 'years of controversy' and have not significantly abated until today – or why GMOs have become a 'sounding board' (Torgersen et al. 2002) for a host of issues ranging from the fear of globalisation and dilution of national identities to the defence of integrated 'agri-cultural' landscapes and familiar (or idealised) food cultures. Clearly, there is no uncontested, objective vantage point from which to adjudicate the issue of agbiotech. Conventional risk–benefit assessments are a poor guide due to scientific complexity and because they take 'little or no account of the social and ethical ramifications of technological systems, including the threats they pose to long-settled patterns of living' (Jasanoff 2006: 288). Given the centrality of public attitudes to relative political acquiescence in the US and resistance in the EU, some have pointed to 'trigger events' and trust gaps to account for the persistence of public concerns. But the cultural context – often neglected because it cannot be easily defined or measured – must also be considered. And it can be combined with political and economic analysis, as done by this book and by the emerging literature on 'cultural political economy' (e.g. Schurman and Munro 2009).

The public relies on pre-existing moral values and cultural identities to make sense of genetic modification. Typically, regulatory trajectories and routines only change once the issue of agbiotech becomes sufficiently salient to mobilise citizens and consumers. Environmental NGOs are crucial actors in this process, but there is no strong evidence that prevailing cultural understandings of GMOs have been successfully fashioned or imposed by political elites or non-state interest groups. In 1992, during a debate in the European Parliament, one parliamentarian called on the public to accept 'new mental images' in the place of traditional concepts of human identity and of nature (Levidow and Carr 1997: 39). Despite support by those committed to a linear view of human progress and ever-growing mastery over the natural world, it appears that traditional conceptions of cultural identity, 'natural' food and agriculture, and moral worldviews have largely endured. Certainly, cultural values and identities are constantly being reproduced through state institutions and the media. But many of these efforts, such as Italy's state-sponsored food education programme – entitled 'Culture that Feeds' (Morgan et al. 2006: 197) – have traditionalist objectives that aim to preserve (or rekindle) rather than modernise societal perceptions of food and agriculture.

The special quality of GMOs lies in their propensity to 'make connections' across the boundaries of culture and nature (Lien 2004: 9), between deeper spiritual or socio-cultural concerns and concrete political debates. The values attached to Europe's traditional landscapes, the cultural associations of agriculture and food, the identity-related dynamics of consumptive practices – they all implicitly express a critical stance on agbiotech which crystallises into a host of popular concerns, often submerged but now brought to the fore. The image of 'Frankenfoods' encapsulates the cultural response of fear and resistance.

In this book, I conceptualised culture as patterns of values, identities, and practices which can be altered but are more amenable to gradual, evolutionary change. The moral and cultural values brought to bear in the agbiotech controversy are subject to historical accumulation, which makes them resistant to dissonant political appeals, PR campaigns, science communication, and media reporting (Zwick 1998: 86–87). In addition, the concept of civilisational dispositions was used to describe even more fundamental, resilient structures of intersubjective meanings – basic cultural outlooks on elementary aspects of the human condition that can be shared by many societies and leave considerable space for internal diversity. In the case of agbiotech, the central disposition concerned historical patterns of the humanity–nature relationship, marked by relative interactivity in Europe and stronger bifurcation in the US. The link between cultural understandings of nature and the debate over GMOs was highlighted by the frequency of European concerns over the 'unnaturalness' of GM food and crops. By analytically unpacking the widespread defence of the 'natural', I drew attention to associations of nature with the spiritually imbued 'natural' order, concerns over authenticity, and the meaning of 'nature as nostalgia' for the protection of cultural traditions. Anxieties about potential physical and environmental risks of agbiotech ('nature as boomerang') were also an important motivation for public opposition to GMOs. Yet, they did not play the pivotal role that was implied by the risk-centric discourses of most commentators, governments, movements, and interest groups.

To make cultural contexts analytically useful, I sought to reconcile the cultural perspective (focused on partly unconscious habits and historical continuity) with an exploration of political dynamics. In this cultural-political approach, discursive entrepreneurs need to interpret the relevant cultural context(s) and try to connect it with their political objectives. This context is not immutable, but cultural change tends to happen over longer time frames, while political action typically aims at short-term results. I demonstrated the relevance of these reflections through an analysis of anti-GMO mobilisation in the US and the EU. In political and economic terms, an alliance of producer groups has so far outspent and largely curtailed the influence of anti-GMO movements in the US, whereas, in Europe, agbiotech has very effectively connected issues of environmentalism, family farming, and consumerism (Ansell et al. 2006; Kurzer and Cooper 2007). This made it possible to construct a broad-based, influential coalition of actors with affiliations right across the political spectrum. But none of the narratives employed by these actors would have had significant political traction without drawing on deep-lying cultural 'sediments'. Thus, apart from a host of well-studied factors – such as political systems, regulatory crises, the media, the structure of the biotech and food industries, and NGOs' capacity for coalition-building – cultural inclinations have played a critical role in the success of anti-GMO campaigns. European citizens have often invoked a fundamentally moral veto regarding 'unnatural' and 'unnecessary' GMOs. Americans have tended to rely on utilitarian criteria of cost–benefit analysis, combining the rationale of economic progress with public health and environmental protection.

Given the more tangible nature of political dynamics and the dominant understanding of culture as a 'thin', residual, and malleable explanatory variable (Mishler and Pollack 2003), I substantiated my argument about deeply rooted values and identities (with relevance for agbiotech) with a long-term historical account of transatlantic cultural divergence. While both continents share a common storyline of increasing exploitation and 'de-spiritualisation' of the natural world, there were growing signs of differentiation early on. America's geographical size and the apparent inexhaustibility of its resources, combined with the crusading spirit of Puritan settlers, led to unprecedented rates of environmental exploitation. The absence of venerated historical signatures in the landscape prepared the ground for the modern forces of rationalisation, industrialisation, and individualisation. Europe's environmental transformation, including the elimination of pagan and animistic spirituality, was much more piecemeal and contested. Europeans' long-running struggle with ecological limits was a recurrent warning against promethean promises.

Chapters 5 and 6 drew on Eder's (Polanyi-inspired) 'double process' of modernity, in which the steady advance of modern rationalisation and commodification is matched by a radicalisation of societal opposition, mainly through recourse to traditions, identity, and nature (Eder 1996). Although, to some extent, this dynamic applies to both world regions, transatlantic differences are crucial. Ever since reactions against industrial modernity began to gather pace in the nineteenth century, the pastoral tradition with its veneration for the humanised, historical, 'cultural' landscape has been the dominant European strand. Most American champions of Romanticism focused on their continent's 'pristine wildernesses' while retaining their faith in technology and slowly abandoning the middle landscape of agriculture. In the US, the idea of nature often expressed an Enlightenment notion of purity and truth. At the same time, in its 'wilderness' manifestation it furnished a spiritual (and materially tangible) symbol of beauty and perfection and acted as 'a vital cohesive force in a country that lacked the glue of ethnic, religious and racial homogeneity' (Coates 1998: 108).

European variants of nature worship encompassed a much greater diversity of meanings and tended to associate it with 'human' constructions, such as historical memory, cultural identity, culinary traditions, food production, and agriculture. The nature–culture dualism was almost dissolved here, for what counted as 'cultural' in southern Europe (e.g. food) was often 'naturalised' in the north. The core insight about 'interactive' European 'cultures of nature' is that they are preoccupied with evidently humanised areas and practices, whereas 'bifurcationist' American society has more successfully separated the human from the natural sphere. This civilisational disposition underlies public reactions to GMOs, insofar as the technology touches upon less salient areas of life in the US, while acting as an ominous 'sounding board' for significant sections of European societies. In many ways, the sceptical or fearful reaction to GMOs by many Europeans is an expression of underlying socio-cultural values and identities. Indeed, some commentators suspect that 'people have found a symbolic arena in which they can put up a resistance to the dynamic force of technology, which has them at its mercy and which is steamrollering society' (van den Daele 2007). Similar responses can at times also be observed in the US, but they are so far limited to counter-cultural movements or to particular areas, such as the coastal regions.

Over time, however, such a revival of profound cultural values and identities might become a broader Western or even global phenomenon. The sociologist Alain Touraine (2004: 129), for instance, has observed that the West is undergoing a 'cultural mutation': '[a]s we question "progress" and discover the absence of transcendent meaning in "history", we rediscover nature.' For him, this necessary project of ecological recalibration is fraught with danger, with pre-Christian temptations of surrendering the human fate to the 'despotism of nature'. But Touraine also acknowledges that the late modern world has yielded an 'empty image of the future' which fuels the search for roots and 'nostalgia for the past' (ibid.: 131). Despite not differentiating between the US and Europe, Touraine's remark provides an elegant summary of the late modern condition in which material and cultural anxieties converge on the defence of the 'natural' as a placeholder for the reformist or reactionary project of an 'alternative modernity'.

The future of agricultural biotechnology

Proponents of agbiotech have come to recognise the depth of public unease (in Europe) and the potential for consumer mobilisation fuelled by demands for the 'right to know' (in the US). Their strategy is no longer to merely portray GMOs as 'natural' or package them as a simplistic narrative of technological progress. The public's desire for a 'value rationality', for a 'moral commitment' (O'Mahony and Skillington 1999: 111), has led agbiotech companies to draw attention to the global South and link GM crops to the objective of ending global hunger in an age of population growth and impending climate change. This narrative is beginning to offer a degree of moral and cultural power, as well as partial empirical veracity, which complements the political influence that agbiotech companies already possess. Over the coming decades, the neo-Malthusian rationale might grow in tandem with genuine scientific advances (especially drought- or salt-tolerant crops) and persuade most citizens that material benefits and a broader form of 'moral usefulness' outweigh the arguments of precaution, of 'natural' or cultural authenticity, and the aspirations of alternative agricultural systems. Over the last five to ten years, the debate (especially in Europe) has clearly expanded to include such broader socio-economic and ethical questions.[1]

An exploration of relevant cultural contexts in both 'world regions' helped to illuminate why in the US, despite some potential regulatory crises, the tailored message of environmental and health risks was insufficient to create widespread public outrage. Most Americans still privilege low price and convenience over paying heed to conceivable, albeit still uncertain, risks. US regulators, backed by influential corporate actors from the 'Food Chain Network', will arguably strive to continue on the path of technological innovation unless substantial new risks are uncovered. An interesting comparison could be drawn with the debate on 'red' (medical) biotechnology. In a mirror image of the transatlantic divide over GMOs, technologies such as stem cell research have attracted considerable moral opprobrium in the US, while Europe has witnessed a weaker and less widespread oppositional movement. For those European countries which did encounter substantial public opposition and passed stringent laws on embryo research, cultural (and especially religious) factors appear to have been decisive (Fink 2008).

Regarding agbiotech, European NGOs were capable of mobilising large sections of European societies by adding moral discourses to the customary stock of (utilitarian) risk-based arguments. Evidently, these campaigns played out differently in individual European countries, and future research should be undertaken to examine national cultural contexts and pluralist interest politics. This would help to establish a more fine-grained typology of national positions on and public attitudes towards agbiotech. Such insights would be especially important at a time when the EU's supranational system of scientific advice and political risk management has entered another reformist phase. In the summer of 2009, a group of eleven member states proposed the partial renationalisation of agbiotech regulation to enable individual countries to legally uphold national bans, even if based on ethical or cultural grounds.[2]

While it might also provoke a renewed transatlantic trade dispute, this kind of reform would certainly increase intra-EU regulatory differences – either through direct restrictions or indirectly through coexistence legislation. Civil society mobilisation and the proliferation of 'GMO-free' regions might induce some countries to use their powers under the 'coexistence' regulations to discourage the planting of GM crops, for instance by imposing stiff penalties for even inadvertent contamination of non-GM fields. A smaller number of pro-GMO governments might, in turn, hope to desensitise consumers by using lax coexistence legislation, thus undercutting the very notion of 'non-GM' products. It remains to be seen whether these countervailing trends will enable GM products to be widely grown and sold alongside non-GM alternatives.

A large comparative study has established that at least some European consumers may be prepared to purchase products with GMO labels (Consumerchoice 2008). While some of these simulated consumer decisions might still change under 'real-world' conditions of civil society mobilisation and discursive framing, a sizeable proportion of consumers would probably be receptive to utilitarian arguments of convenience, price, and potential health benefits. Whereas I have emphasised a 'top-down' process of the attitude formation with regard to GMOs (Grunert et al. 2003), which refers to the centrality of higher order values and attitudes, others have made the case for more conventional benefit perception through a 'bottom-up', product-based process of judgement (Gaskell et al. 2004). Especially northern European consumers may indeed be susceptible to utilitarian benefits, but the cultural values and identities associated with food and agriculture mean that Europeans' cognitive threshold is higher, on average, than that of US consumers. As agbiotech supporters have begun to frame their innovations in ethical terms, they may succeed in engaging European societies more positively. This adapted message will not resonate at an equally deep level of cultural identity, but it may help to re-brand agbiotech as an essential element of humanistic 'progress', thus assuaging opposition to the modification of 'nature' and 'heritage'. Without a strong (ethical or utilitarian) rationale for agbiotech products, the culturally motivated resistance of many Europeans is unlikely to weaken significantly.

Amid a renewed wave of anti-GMO mobilisation across many US states – centred on the minimalist goal of sovereign consumers' 'right to know' – regulatory change across the Atlantic is also on the cards. Recent decisions (in May 2014) in favour of mandatory labelling in Vermont and in two Oregonian counties show that the political and structural power of the agbiotech alliance can be overcome, at least in some US regions. A new federal bill to pre-empt mandatory labelling and enshrine voluntary labels may yet be passed, but the unsettled coexistence of non-GM, conventional, and organic agriculture will continue to pose significant challenges. Any move to commercialise genetically modified animals (e.g. GM salmon) will likely raise the level of public awareness and concern. In general, the US agbiotech industry would be well advised to look beyond agronomic efficiency and utilitarian discourses when engaging with the public. As demonstrated in a qualitative study of US consumer attitudes, functional benefits (such as health, taste, nutrition) and technological curiosity were reasons to endorse GM products, but they were often understood in an emotional and situated rather than in an abstract and economistic way. Kniazeva (2005: 35–36) concludes that if GM products were presented as an exciting personal 'problem-solver', US marketing strategies might overcome the current lukewarm approval and help foster a more heart-felt public acceptance.

When Prakash and Kollman (2003) first suggested that the US may soon begin to emulate the EU's more cautious strategy and introduce policies on crop segregation, traceability, and labelling, their prognosis was premature. But they may yet prove their prescience if the pressures for reform in both the US and the EU manage to inflect current regulatory trajectories. Just like the forecasts offered by other scholars, the cultural-political approach I have adopted in this book cannot hope to accurately predict the future of agbiotech in the US and the EU. But the importance of cultural values, public opinion, and consumer behaviour suggests that, in the absence of major global crises (such as sharply rising food prices or rapid climate change), the European fortunes of agbiotech will remain uncertain. In the US, on the other hand, further regulatory crises and demonstrable harm might be required before public mobilisation will genuinely drive political decision-making. Nevertheless, the widely shared ideal of 'consumer sovereignty' and the potential spread of mandatory labelling of GMOs could become a serious threat to current regulatory apathy.

Notes

Introduction

1. StarLink was a GM corn variety in the US which, although only permitted for use in animal feed, was detected in taco shells.
2. However, US consumers were significantly more likely to acknowledge potential benefits.

1 Overview of Regulatory Frameworks and Public Opinion

1. Figures from September 2010 show that 169 regions, 123 provinces, and 4713 local authorities have declared themselves 'GMO-free': see http://bit.ly/1aYa6rJ (accessed 07 May 2014).

2 Perspectives on Regulatory Divergence

1. While the latter is undoubtedly true, more recent legislative activity has often been associated with bills and referendums seeking the mandatory labelling of GM products.
2. The Senior Advisory Group on Biotechnology (SAGB) was created by around 35 leading biotechnology corporations, including Monsanto, Novartis, Syngenta, Rhone-Poulenc, AgrEvo, Aventis, and Zeneca. It merged with the Secretariat of National Bio-Industry Associations (ESN) – composed of around 500 smaller biotech companies – in 1996 to form EuropaBio.
3. Ulrich Beck (1992) argues that the era of modernity and its technological innovations have created new and often unknown kinds of risks which will come to preoccupy modern societies as they move from 'simple modernity' towards 'reflexive modernisation'.
4. European opinions from the year 2000 show that only 30% of respondents believed that industry 'did good work for society', but 70% said this about consumer organisations (Tsioumani 2004).

3 Theorising Culture and Nature

1. The concept put forward here differs from two established schools of cultural materialism. The anthropological-materialist strand (Harris 1968) seeks to apply the tools of positivist social science, whereas the cultural studies strand (e.g. Wilson 1995) focuses on the ability of socio-political elites to shape popular culture to stabilise hegemonic ideas and social practices.
2. In later chapters, I will occasionally use civilisation in precisely this sense to denote a binary opposition to raw 'wilderness'.

3. Braudel views the concept of *longue durée* as a crucial innovation and distinguishes it from both traditional events-based history and the history of periods, epochs and, cycles.

4 Cultural Politics and Resistance to GMOs

1. As Chapter 5 will show, some instances of 'nature' have considerable cultural appeal in the US: wilderness is the foremost example.
2. Ultimately, the total number of participating NGOs reached 108.
3. See http://www.sierraclub.org/biotech/report.aspx (accessed 12 May 2014).
4. See http://www.nclnet.org/about-ncl/policy-statements (accessed 12 May 2014).
5. FDA's period for public comment ended in May 2013 and attracted more than 1.8 million responses.
6. See http://www.nongmoproject.org/2013/09/17/non-gmo-project-moves-to-expand-verification-capabilities/ (accessed 12 May 2014).
7. The industry's new website for countering activist campaigns is thus called http://gmoanswers.com/.
8. The scientist Árpád Pusztai publicly announced that his studies had shown that rats fed on a diet of GM potatoes suffered serious health impacts. Soon after, his employer, the Rowett Research Institute in Aberdeen, repudiated his findings and terminated his contract.
9. In the UK, for instance, this coalition spanned a wide political spectrum, ranging from radical green group Earth First! to the traditionalist Townswomen's Guild.
10. As a specialised agri-business corporation with a vital interest in the global success of agbiotech, an advertisement campaign was a logical strategic move. However, it was badly timed and relied on a tone of confident superiority. Under the headline 'Food biotechnology is a matter of opinion, Monsanto believes you should hear them all,' the company presented a new benefit of GMOs every weekend to broadsheet readers in Britain. Activists retorted with 'Food biotechnology is a matter of opinion. At Nonsanto, opinion is something we buy. Nonsanto, Fraud – Stealth – Hype'.
11. Similar dynamics might also apply to the US. The three states which had passed labelling bills (at the time of writing) all have a comparatively high number of organic farms.
12. One may speculate on whether this form of 'latent' mobilisation might explain certain political decisions, such as the Spanish government's frequent abstentions on GMO authorisations at the EU level. In the US, on the other hand, consistently high scores for mandatory labelling in public surveys indicate latency, but have only recently become more politically salient.

5 Environmental History: Nature, Landscapes, and Identities

1. 'New Founde Land', an expression in the old English language, survives in the name of the region of Newfoundland.

2. Civilisation here refers to the conventional meaning of refinement and progress, often used in the context of American environmental history and, occasionally in this book, when emphasising the separation of wilderness/nature and humanity.
3. The complete paper is available online at http://xroads.virginia.edu/~HYPER/TURNER/chapter1.html (accessed 13 May 2014).
4. The Mediterranean population doubled twice: during 600–1000 and once more during 1000–1200 (Hughes 2005: 71).
5. Another historical comparison confirms the 'glacial' pace of Europe's environmental transformation: whereas Europe's rate of environmental change stands at 0.1–0.3% per year over the last fifteen centuries, America boasts an impressive 0.7% (1600–1900), which is still dwarfed by today's tropical rainforests (1%) (Dobson cited in Whited et al. 2005: 85).
6. The alternative terminology of 'Renaissance' reveals a more optimistic inclination and begins around 100 years earlier. Given this minor distinction, I use the terms interchangeably.
7. One critic, Dr Adam Seybert, was adamant that even the unpleasant aspects of nature, such as swamps, were an integral part of it and likely served some higher purpose (Glacken 1967: 689).

6 Agri-Cultural and Culinary Identities

1. Narrowly defined environmental organisations only assemble around 1.5% of the British population (Dalton 2005), but the broader definition used here (including heritage, landscape, tradition, etc.) exponentially increases membership figures and, by extension, the number of sympathisers. The National Trust alone counted more than 5% of the population among its members in 2004.
2. This also explains why organic agriculture, with its emphasis on natural processes rather than artisanal quality, is still a relatively niche market in France.
3. The impact of what 'happens' is determined by the *moral* rather than *material* quality of change. In the case of Scottish moors, this includes substantial physical alterations. In the case of agbiotech, this refers to the moral debasing of agriculture through its 'hyper-technological' turn – much more so than first green revolution with its fertilisers and pesticides. Over time, like in Italy, organic farming in France might yet be re-framed as the 'true' form of agriculture. For now, conventional industrial agriculture is tolerated, but not celebrated by the majority of the public.
4. Americans call the Germans *Krauts* and the British sneer at the French *frogs*, whereas the French have dubbed the British *les roastbeefs* and the Italians *les macaronis*.
5. Some European countries, especially in the south, fit the model of traditionalist resistance better than others. Britain, for example, has generally followed a modern industrialist trajectory, but has also seen the recovery or re-invention of food culture during the last 20 years.

Conclusion

1. Moreover, the rationale of utilitarian benefit is being promoted by messages of health and longevity through GM innovations. In autumn 2008, scientists from the UK fed GM tomatoes (rich in antioxidant pigments normally found in the snapdragon flower) to laboratory mice and noted positive health effects. They declared that this was 'certainly the first example of a GMO with a trait that really offers a potential benefit for all consumers' (BBC 2008). Another group of UK scientists began a similarly promising crop trial in spring 2014, seeking to produce fish oils (containing healthy omega-3 fatty acids) in the seeds of GM camelina plants.
2. This group included ten GM-sceptical nations (Austria, Bulgaria, Cyprus, Greece, Hungary, Ireland, Latvia, Lithuania, Malta, and Slovenia) and the pro-GMO Netherlands.

Bibliography

Agar, N. (2003) 'GM Food, Risk, and Taste', *Biology and Philosophy*, 18, pp. 599–606.

Agnoletti, M. (2003) 'Historical Research and Landscape Analysis: The Case of Tuscany'. Paper presented at the 2nd International Conference of the European Society for Environmental History (ESEH), Prague, Czech Republic, 3–7 September.

Albrecht, G., G.-M. Sartore, L. Connor, N. Higginbotham, S. Freeman, B. Kelly, H. Stain, A. Tonna and G. Pollard. (2007) 'Solastalgia: The Distress Caused by Environmental Change', *Australasian Psychiatry*, 15(S1), pp. S95–S98.

Alemanno, A. (2006) 'Food Safety and the Single European Market', in C. Ansell and D. Vogel (eds.) *What's the Beef? The Contested Governance of European Food Safety* (Cambridge, MA: MIT Press), pp. 237–258.

Anderson, J. C., C. J. Wachenheim and W. C. Lesch. (2006) 'Perceptions of Genetically Modified and Organic Foods and Processes', *AgBioForum*, 9(3), pp. 180–194.

Andrée, P. (2005) 'The Genetic Engineering Revolution in Agriculture and Food: Strategies of the "Biotech Bloc"', in D. L. Levy and P. J. Newell (eds.) *The Business of Global Environmental Governance* (Cambridge, MA: MIT Press), pp. 135–166.

——— (2007) *Genetically-Modified Diplomacy: The Global Politics of Agricultural Biotechnology and the Environment* (Vancouver: University of British Columbia Press).

Ansell, C. (2006) 'The Asymmetries of Governance', in C. Ansell and D. Vogel (eds.) *What's the Beef? The Contested Governance of European Food Safety* (Cambridge, MA: MIT Press), pp. 329–349.

Ansell, C., R. Maxwell and D. Sicurelli. (2006) 'Protesting Food: NGOs and Political Mobilization in Europe', in C. Ansell and D. Vogel (eds.) *What's the Beef? The Contested Governance of European Food Safety* (Cambridge, MA: MIT Press), pp. 97–122.

Aquino, H. K. and W. K. Hallman. (2004). 'Differences in Perceptions of Agricultural Biotechnology: A Comparison Study Between Germany and the United States'. Paper presented at the Conference 'Scientific Knowledge and Cultural Diversity', Barcelona, Spain, 3–6 June.

Areal, F., L. Riesgo and E. Rodriguez-Cerezo. (2011) 'Attitudes of European Farmers Towards GM Crop Adoption', *Plant Biotechnology Journal*, 9, pp. 945–957.

Arnold, D. (1996) *The Problem of Nature: Environment, Culture and European Expansion* (Oxford: Blackwell).

Bailey, R. (2005) *Liberation Biology: The Scientific and Moral Case for the Biotech Revolution* (Amherst, NY: Prometheus Books).

Baker, A. R. H. (2003) *Geography and History: Bridging the Divide* (Cambridge: Cambridge University Press).

Baldwin, P. (2009) *The Narcissism of Minor Differences: How America and Europe Are Alike* (Oxford: Oxford University Press).

Barinaga, M. (2000) 'Asilomar Revisited: Lessons for Today?' *Science*, 287(5458), pp. 1584–1585.
Barrows, G., S. Sexton and D. Zilberman. (2014) 'Agricultural Biotechnology: The Promise and Prospects of Genetically Modified Crops', *The Journal of Economic Perspectives*, 28(1), pp. 99–119.
Barry, J., G. Ellis and C. Robinson. (2008) 'Cool Rationalities and Hot Air: A Rhetorical Approach to Understanding Debates on Renewable Energy', *Global Environmental Politics*, 8(2), pp. 67–98.
Baudrillard, J. (1988) *America* (London: Verso).
Bauer, M. W., J. Durant and G. Gaskell. (1997) 'Europe Ambivalent on Biotechnology', *Nature*, 387(6636), pp. 845–847.
Bauer, M. W., M. Kohring, A. Allansdottir and J. Gutteling. (2001) 'The Dramatisation of Biotechnology in Elite Mass Media', in G. Gaskell and M. W. Bauer (eds.) *Biotechnology 1996–2000: The Years of Controversy* (London: Science Museum), pp. 35–52.
Baylis, K., S. Peplow, G. Rausser and L. Simon. (2008) 'Agri-environmental Policies in the EU and United States: A Comparison', *Ecological Economics*, 65(4), pp. 753–764.
BBC. (2008) 'Purple Tomato "may boost health" '. 26 October. Available at http://news.bbc.co.uk/go/pr/fr/-/1/hi/health/7688310.stm (accessed 30/05/2014).
Beck, U. (1992) *Risk Society: Towards a New Modernity* (London: Sage).
Becker, G. S. and T. Cowan. (2006) *Agricultural Biotechnology: Background and Recent Issues*. Congressional Research Service, Order Code RL32809.
Beekman, V. (2006) 'Feeling Food: The Rationality of Perception', *Journal of Agricultural and Environmental Ethics*, 19, pp. 301–312.
Beinart, W. and P. Coates. (1995) *Environment and History: The Taming of Nature in the USA and South Africa* (London: Routledge).
Belson, N. A. (2000) 'US Regulation of Agricultural Biotechnology: An Overview', *AgBioForum*, 3(4), pp. 268–280.
Benford, R. D. and D. A. Snow. (2000) 'Framing Processes and Social Movements: An Overview and Assessment', *Annual Review of Sociology*, 26, pp. 611–639.
Benhabib, S. (2002) *The Claims of Culture: Equality and Diversity in the Global Era* (Princeton: Princeton University Press).
Bernauer, T. (2003) *Genes, Trade, and Regulation: The Seeds of Conflict in Food Biotechnology* (Princeton: Princeton University Press).
Bernauer, T. and E. Meins. (2003) 'Technological Revolution Meets Policy and the Market: Explaining Cross-national Differences in Agricultural Biotechnology Regulation', *European Journal of Political Research*, 42(5), pp. 643–683.
Berry, W. (1977) *The Agricultural Crisis: A Crisis of Culture* (New York: The Myrin Institute).
Bess, M. (2003) *The Light-Green Society: Ecology and Technological Modernity in France, 1960–2000* (Chicago: University of Chicago Press).
Bevilacqua, D. (2007) 'The International Regulation of Genetically Modified Organisms: Uncertainty, Fragmentation, and Precaution', *European Environmental Law Review*, 16(12), pp. 314–336.
Bishop, P. (1991) 'Constable Country: Diet, Landscape and National Identity', *Landscape Research*, 16(2), pp. 31–36.
Black, R. (2008) 'Missing the Signs of Genetic Irrelevance'. *BBC News*, 13 August. Available at http://news.bbc.co.uk/go/pr/fr/-/1/hi/sci/tech/7558585.stm (accessed 17/05/2014).

Bloch, M. (1966) *French Rural History: An Essay on Its Basic Characteristics* (London: Routledge & Kegan Paul).

Bolt, C. and A. R. Lee. (1989) 'New England in the Nation', in M. Bradbury and H. Temperley (eds.) *Introduction to American Studies*, 2nd edition (London: Longman), pp. 78–106.

Bonny, S. (2003) 'Why Are Most Europeans Opposed to GMOs? Factors Explaining Rejection in France and Europe', *Electronic Journal of Biotechnology*, 6(1), pp. 50–71. Available at http://www.ejbiotechnology.info/content/vol6/issue1/full/4.

Bradbury, M. and H. Temperley. (1989) 'Introduction', in M. Bradbury and H. Temperley (eds.) *Introduction to American Studies*, 2nd edition (London: Longman), pp. 1–30.

Braudel, F. (1995) [1963] *A History of Civilizations* (Harmondsworth: Penguin).

Bray, P. M. (1998) 'Italian Park and Protected Areas: Experience and Twinning', *The George Wright Forum*, 15(2), pp. 20–23.

Bredahl, L. (2001) 'Determinants of Consumer Attitudes and Purchase Intentions with Regard to Genetically Modified Foods', *Journal of Consumer Policy*, 24, pp. 23–61.

Bretherton, C. and J. Vogler. (2006) *The European Union as a Global Actor*, 2nd edition (Abingdon: Routledge).

Brookes, G. and P. Barfoot. (2014) 'Economic Impact of GM Crops: The Global Income and Production Effects 1996–2012', *GM Crops and Food*, 5(1), pp. 37–47.

Brown, W. (2011) 'It's Time for a New Biotechnology Law'. The Brookings Institution, 29 July. Available at http://www.brookings.edu/opinions/2011/0727_biotechnology_law_brown.aspx (accessed 04/05/2014).

Brulle, R. J. and J. C. Jenkins. (2006) 'Spinning Our Way to Sustainability?' *Organization & Environment*, 19(1), pp. 82–87.

Buijs, A. E., B. J. M. Arts, B. H. M. Elands and J. Lengkeek. (2011) 'Beyond Environmental Frames: The Social Representation and Cultural Resonance of Nature in Conflicts over a Dutch Woodland', *Geoforum*, 42(3), pp. 329–341.

Bunce, M. (1994) *The Countryside Ideal: Anglo-American Images of Landscape* (London: Routledge).

Burchell, R. A. and R. J. Gray. (1989) 'The Frontier West', in M. Bradbury and H. Temperley (eds.) *Introduction to American Studies*, 2nd edition (London: Longman), pp. 130–156.

Burns, C. (2012) 'How and When Did We Get Here? An Historical Institutionalist Analysis of EU Biotechnology Policy', *Journal of European Integration*, 34(4), pp. 341–357.

Buttel, F. H. (2005) 'The Environmental and Post-Environmental Politics of Genetically Modified Crops and Foods', *Environmental Politics*, 14(3), pp. 309–323.

Carrell, S. (2008) '£500m Project Offers Jobs and Income, but Will It Devastate the Environment?' *The Guardian*, 04 February. Available at http://www.theguardian.com/environment/2008/feb/04/windpower.renewableenergy (accessed 12/08/2014).

Carter, N. (2007) *The Politics of the Environment: Ideas, Activism, Policy*, 2nd edition (Cambridge: Cambridge University Press).

Caulder, J. (1998) 'Agricultural Biotechnology and Public Perceptions', *AgBioForum*, 1(1), pp. 38–39.

Ceccoli, S. and W. Hixon. (2012) 'Explaining Attitudes Toward Genetically Modified Foods in the European Union', *International Political Science Review*, 33(3), pp. 301–319.

Chaney, S. (2005) 'Protecting Nature in a Divided Nation: Conservation in the Two Germanys, 1945–1972', in T. Lekan and T. Zeller (eds.) *Germany's Nature: Cultural Landscapes and Environmental History* (New Brunswick, NJ: Rutgers University Press), pp. 207–243.

Chapman, A. (2005) 'Genetic Engineering: The Unnatural Argument', *Techné: Research in Philosophy and Technology*, 9(2), pp. 81–93.

Charlton, D. G. (1984) *New Images of the Natural in France: A Study in European Cultural History 1750–1800* (Cambridge: Cambridge University Press).

Cheyne, I. (2008) 'Life After the Biotech Products Dispute', *Environmental Law Review*, 10(1), pp. 52–64.

Chiarabolli, A. (2011) 'Coexistence of Genetically Modified Crops with Conventional and Organic Agriculture: The Italian Situation', *Biotechnology Law Report*, 30(6), pp. 675–683.

Clapp, J. and D. Fuchs. (2009) 'Agrifood Corporations, Global Governance, and Sustainability: A Framework for Analysis', in J. Clapp and D. Fuchs (eds.) *Corporate Power in Global Agrifood Governance* (Cambridge, MA: MIT Press), pp. 1–25.

Clout, H. (1998) 'Rural Europe Since 1500: Areas of Innovation and Change', in R. A. Butlin and R. A. Dodgshon (eds.) *An Historical Geography of Europe* (Oxford: Clarendon Press), pp. 225–242.

Coates, P. (1998) *Nature: Western Attitudes Since Ancient Times* (Cambridge: Polity Press).

Coburn, C. V. (2005) 'Out of the Petri Dish and Back to the People: A Cultural Approach to GMO Policy', *Wisconsin International Law Journal*, 23, pp. 283–320.

Coghlan, A. (2003) 'UK Public Strongly Rejects GM Foods'. *New Scientist*, 24 September. Available at http://www.newscientist.com/article/dn4191#.UekeJtPTUiI (accessed 02/05/2014).

Collins, H. and T. Pinch. (1998) *The Golem – What You Should Know About Science* (Cambridge: Cambridge University Press).

Commandeur, P., P.-B. Joly, L. Levidow, B. Tappeser and B. Terragni. (1996) 'Public Debate and Regulation of Biotechnology in Europe', *Biotechnology and Development Monitor*, 26, pp. 2–9.

Connor, M. and M. Siegrist. (2010) 'Factors Influencing People's Acceptance of Gene Technology: The Role of Knowledge, Health Expectations, Naturalness, and Social Trust', *Science Communication*, 32(4), pp. 514–538.

Consumerchoice. (2008) *Do European Consumers Buy GM Foods?* Research Project funded by the European Commission (Framework 6). Available at http://www.kcl.ac.uk/medicine/research/divisions/dns/projects/consumerchoice/index.aspx (accessed 30/05/2014).

Cooley, D. R. and G. A. Goreham. (2004) 'Are Transgenic Organisms Unnatural?' *Ethics & The Environment*, 9(1), pp. 46–55.

Cormick, C. (2005) 'Lies, Deep Fries, and Statistics!! The Search for the Truth Between Public Attitudes and Public Behaviour Towards Genetically Modified Foods', *Choices*, 20(4), pp. 227–231.

Costa-Font, M. and J. M. Gil. (2012) 'Meta-attitudes and the Local Formation of Consumer Judgments Towards Genetically Modified Food', *British Food Journal*, 114(10), pp. 1463–1485.

Cowan, T. (2010) *Agricultural Biotechnology: Background and Recent Issues*. Order Code RL32809. Congressional Research Service.

Cox, A. (2007) 'Chimeraphobia', *The Philosophers' Magazine*, 38, pp. 14–16.

Cox, R. W. (2002) *The Political Economy of a Plural World: Critical Reflections on Power, Morals, and Civilization* (London: Routledge).

Cronon, W. (1996) 'The Trouble with Wilderness: Or, Getting Back to the Wrong Nature', in W. Cronon (ed.) *Uncommon Ground: Rethinking the Human Place in Nature* (New York: W. W. Norton), pp. 69–90.

Crossley, N. (2002) *Making Sense of Social Movements* (Buckingham: Open University Press).

Dake, K. (1992) 'Myths of Nature: Culture and the Social Construction of Risk', *Journal of Social Issues*, 48(4), pp. 21–37.

Dalton, R. J. (2005) 'The Greening of the Globe? Cross-National Levels of Environmental Group Membership', *Environmental Politics*, 14(4), pp. 441–459.

Daniels, S. (1993) *Fields of Vision: Landscape Imagery and National Identity in England and the United States* (Cambridge: Polity Press).

Davison, A., I. Barns and R. Schibeci. (1997) 'Problematic Publics: A Critical Review of Surveys of Public Attitudes to Biotechnology', *Science, Technology & Human Values*, 22(3), pp. 317–348.

Daviter, F. (2009) 'Schattschneider in Brussels: How Policy Conflict Reshaped the Biotechnology Agenda in the European Union', *West European Politics*, 32(6), pp. 1118–1139.

Deane-Drummond, C., R. Grove-White and B. Szerszynski. (2003) 'Genetically Modified Theology: The Religious Dimensions of Public Concerns About Agricultural Biotechnology', in C. Deane-Drummond, B. Szerszynski and R. Grove-White (eds.) *Re-Ordering Nature: Theology, Society and the New Genetics* (London: T & T Clark), pp. 17–38.

Deléage, J. P. and D. Hémery. (1990) 'From Ecological History to World Ecology', in P. Brimblecombe and C. Pfister (eds.) *The Silent Countdown: Essays in European Environmental History* (Berlin: Springer-Verlag), pp. 21–36.

Dobbs, M. (2011) 'Excluding Coexistence of GMOs? The Impact of the EU Commission's 2010 Recommendation on Coexistence', *RECIEL*, 20(2), pp. 180–193.

Doherty, B. and G. Hayes. (2012) 'Tactics, Traditions and Opportunities: British and French Crop-trashing Actions in Comparative Perspective', *European Journal of Political Research*, 51(4), pp. 540–562.

Dona, A. and I. S. Arvanitoyannis. (2008) 'Health Risks of Genetically Modified Foods', *Critical Reviews in Food Science and Nutrition*, 49(2), pp. 164–175.

Donahue, B. (2004) *The Great Meadow: Farmers and the Land in Colonial Concord* (New Haven: Yale University Press).

Douglas, M. (1975) *Implicit Meanings: Essays in Anthropology* (London: Routledge & Kegan Paul).

Douglas, M. and A. Wildavsky. (1982) *Risk and Culture: An Essay on the Selection of Technical and Environmental Dangers* (Berkeley: University of California Press).

Dower, J. (2001) 'The Variation in Landscape Policies Around Europe', in ICOMOS-UK (eds.) *The Cultural Landscape: Planning for a Sustainable Partnership*

Between People and Place (London: International Council on Monuments and Sites UK), pp. 96–108.

Dreezens, E., C. Martijn, P. Tenbült, G. Kok and N. K. de Vries. (2005) 'Food and Values: An Examination of Values Underlying Attitudes Toward Genetically Modified- and Organically Grown Food Products', *Appetite*, 44(1), pp. 115–122.

Drezner, D. W. (2007) *All Politics Is Global: Explaining International Regulatory Regimes* (Princeton: Princeton University Press).

Dunin-Wasowicz, T. (1990) 'Natural Environment and Human Settlement over the Central European Lowland in the 13th Century', in P. Brimblecombe and C. Pfister (eds.) *The Silent Countdown: Essays in European Environmental History* (Berlin: Springer-Verlag), pp. 92–105.

Dyer, H. C. (1996) 'Environmental Security as a Universal Value: Implications for International Theory', in J. Vogler and M. F. Imber (eds.) *The Environment and International Relations* (London: Routledge), pp. 22–40.

Eagleton, T. (2000) *The Idea of Culture* (Oxford: Blackwell).

Earle, T. C. and G. T. Cvetkovich. (1995) *Social Trust: Toward a Cosmopolitan Society* (Westport: Praeger).

Eccleston, P. (2007) 'Intensive Farming Damaging Bird Numbers', *The Daily Telegraph*, 6 June. Available at http://www.telegraph.co.uk/earth/earthnews/3296537/Intensive-farming-damaging-bird-numbers.html (accessed 12/08/2014).

Echols, M. A. (1998) 'Food Safety Regulation in the European Union and the United States: Different Cultures, Different Laws', *Columbia Journal of European Law*, 4, pp. 525–543.

Eder, K. (1996) *The Social Construction of Nature: A Sociology of Ecological Enlightenment* (London: Sage).

Edge, J. L. (2011) *Strategizing Beyond the State: The Global Environmental Movement and Corporate Actors*. Hamilton, McMaster University. PhD Thesis. Available at http://digitalcommons.mcmaster.ca/opendissertations/6416/ (accessed 12/05/2014).

Elias, N. (1978) *The Civilizing Process* (Oxford: Blackwell).

Elkington, J. and T. Burke. (1987) *The Green Capitalists: Making Business Sense of the Environment* (London: Gollancz).

Euractiv (2013) 'Food Safety Agency Faces Fresh Accusation of Industry Ties'. 29 October. Available at http://www.euractiv.com/health/eu-food-safety-agency-accused-cl-news-531366 (accessed 06/05/2014).

European Federation of Biotechnology (EFB). (1999) *Ethical Aspects of Agricultural Biotechnology*: Task Group on Public Perceptions of Biotechnology.

Ezzell, C. (1987) 'US Attitudes to Biotechnology Show Qualified Support', *Nature*, 327(11 June), pp. 453–453.

Falkner, R. (2007) 'The Global Biotech Food Fight: Why the United States Got It So Wrong', *Brown Journal of World Affairs*, 14(1), pp. 99–110.

——— (2008) *Business Power and Conflict in International Environmental Politics* (Houndmills: Palgrave).

——— (2009) 'The Troubled Birth of the "Biotech Century": Global Corporate Power and Its Limits', in J. Clapp and D. Fuchs (eds.) *Corporate Power in Global Agrifood Governance* (Cambridge, MA: MIT Press), pp. 225–251.

Feindt, P. H. and A. Oels. (2005) 'Does Discourse Matter? Discourse Analysis in Environmental Policy Making', *Journal of Environmental Policy & Planning*, 7(3), pp. 161–173.

Fieldhouse, P. (1986) *Food & Nutrition: Customs & Culture* (London: Croom Helm).

Fink, S. (2008) 'Politics as Usual or Bringing Religion Back in? The Influence of Parties, Institutions, Economic Interests, and Religion on Embryo Research Laws', *Comparative Political Studies*, 41(12), 1631–1656.

Fink, W. and M. Rodemeyer. (2007) 'Genetically Modified Foods: US Public Opinion Research Polls', in D. Brossard, J. Shanahan and T. C. Nesbitt (eds.) *The Public, the Media and Agricultural Biotechnology* (Wallingford: CABI), pp. 126–160.

Fischler, C. (1988) 'Food, Self and Identity', *Social Science Information*, 27(2), pp. 275–292.

Fitzgerald, R. and H. Campbell. (2001) 'Food Scares and GM: Movement on the Nature/Culture Fault Line', *Australian Review of Public Affairs*, 5 October. Available at http://www.australianreview.net/digest/2001/10/fitzgerald_campbell.html (accessed 10/05/2014).

Flammang, J. A. (2009) *The Taste for Civilization: Food, Politics, and Civil Society* (Urbana: University of Illinois Press).

Food Ethics Council. (1999) *Novel Foods: Beyond Nuffield.* Report by the Independent Council for Ethical Standards in Food and Agriculture. Available at http://www.foodethicscouncil.org/system/files/novelfoods.pdf (accessed 16/05/2014).

Fowler, P. (2001) 'Cultural Landscapes: Great Concept, Pity About the Phrase', in ICOMOS-UK (ed.) *The Cultural Landscape: Planning for a Sustainable Partnership Between People and Place* (London: International Council on Monuments and Sites UK), pp. 64–82.

Frewer, L. J., J. Scholderer and L. Bredahl. (2003) 'Communicating About the Risks and Benefits of Genetically Modified Foods: The Mediating Role of Trust', *Risk Analysis*, 23(6), pp. 1117–1133.

Friedman, J. (1994) *Cultural Identity and Global Process* (London: Sage).

Gambino, R. (1998) 'Parks and Protected Areas in Italy: An Overview', *The George Wright Forum*, 15(2), pp. 30–42.

Gamson, W. A. and D. S. Meyer. (1996) 'Framing Political Opportunity', in D. McAdam, J. D. McCarthy and M. N. Zald (eds.) *Comparative Perspectives on Social Movements: Political Opportunities, Mobilizing Structures, and Cultural Framings* (Cambridge: Cambridge University Press), pp. 275–290.

Ganière, P., W. S. Chern and D. Hahn. (2006) 'A Continuum of Consumer Attitudes Toward Genetically Modified Foods in the United States', *Journal of Agricultural Resource Economics*, 31(1), pp. 129–149.

Gaskell, G. and M. W. Bauer. (2001) 'Biotechnology in the Years of Controversy: A Social Scientific Perspective', in G. Gaskell and M. W. Bauer (eds.) *Biotechnology 1996–2000: The Years of Controversy* (London: Science Museum), pp. 3–11.

Gaskell, G., P. Thompson and N. Allum. (2002) 'Worlds Apart? Public Opinion in Europe and the USA', in M. W. Bauer and G. Gaskell (eds.) *Biotechnology: The Making of a Global Controversy* (Cambridge: Cambridge University Press), pp. 351–378.

Gaskell, G., E. Einsiedel, S. Priest, T. Ten Eyck, N. Allum and H. Torgersen. (2001) 'Troubled Waters: The Atlantic Divide on Biotechnology Policy', in G. Gaskell and M. W. Bauer (eds.) *Biotechnology 1996–2000: The Years of Controversy* (London: Science Museum), pp. 96–115.

Gaskell, G., N. Allum, W. Wagner, N. Kronberger, H. Torgersen, J. Hampel and J. Bardes. (2004) 'GM Foods and the Misperception of Risk Perception', *Risk Analysis*, 24(1), pp. 185–194.

Gaskell, G., N. Allum, W. Wagner, T. H. Nielsen, E. Jelsøe, M. Kohring and M. Bauer. (2001) 'In the Public Eye: Representations of Biotechnology in Europe', in G. Gaskell and M. W. Bauer (eds.) *Biotechnology 1996–2000: The Years of Controversy* (London: Science Museum), pp. 53–79.

Gaskell, G., S. Stares, A. Allansdottir, N. Allum, P. Castro, Y. Esmer, C. Fischler, J. Hampel, J. Jackson, N. Kronberger, N. Mejlgaard, A. Quintanilha, A. Rammer, G. Revuelta, P. Stoneman, H. Torgersen and W. Wagner. (2010) *Europeans and Biotechnology in 2010: Winds of Change*. Report to the European Commission's Directorate-General for Research.

Gaudillière, J.-P. (2006) 'Globalization and Regulation in the Biotech World: The Transatlantic Debates over Cancer Genes and Genetically Modified Crops', *OSIRIS*, 21, pp. 251–272.

Geertz, C. (1993) *The Interpretation of Cultures: Selected Essays* (New York: Basic Books).

Gillam, C. (2010) 'Special Report: Are Regulators Dropping the Ball on Biocrops?'. *Reuters*, 13 April. Available at http://www.reuters.com/article/2010/04/13/us-usa-gmos-regulators-idUSTRE63C2AJ20100413 (accessed 04/05/2014).

Gillam, C. and C. Doering. (2011) 'U.S. Farmers Get Approval to Plant GMO Alfalfa'. *Reuters*, 27 January. Available at http://www.reuters.com/article/2011/01/27/gmo-alfalfa-usda-idUSN2727513020110127 (accessed 25/05/2014).

Glacken, C. J. (1967) *Traces on the Rhodian Shore: Nature and Culture in Western Thought From Ancient Times to the End of the Eighteenth Century* (Berkeley, CA: University of California Press).

GMO-Compass. (2007) 'Farmers in the UK and in Germany Are Open to GM Crops'. Available at http://www.gmo-compass.org/eng/news/messages/200705.docu.html (accessed 08 May 2014).

Gordon, P. H. and N. de Boisgrollier. (2005) 'Why the French Love Their Farmers'. *YaleGlobal Online*. Available at http://yaleglobal.yale.edu/content/why-french-love-their-farmers (accessed 16/05/2014).

Gottweis, H. (1998) *Governing Molecules: The Discursive Politics of Genetic Engineering in Europe and the United States* (Cambridge, MA: MIT Press).

Grabner, P., J. Hampel, N. Lindsey and H. Torgersen. (2001) 'Biopolitical Diversity: The Challenge of Multilevel Policy-making', in G. Gaskell and M. W. Bauer (eds.) *Biotechnology 1996–2000: The Years of Controversy* (London: Science Museum), pp. 15–34.

Graff, G. D., G. Hochman and D. Zilberman. (2009) 'The Political Economy of Agricultural Biotechnology Policies', *AgBioForum*, 12(1), pp. 34–46.

Greener, I. (2005) 'The Potential of Path Dependence in Political Studies', *Politics*, 25(1), pp. 62–72.

Grunert, K. G., L. Bredahl and J. Scholderer. (2003) 'Four Questions on European Consumers' Attitudes Towards the Use of Genetic Modification in Food Production', *Innovative Food Science and Emerging Technologies*, 4, pp. 435–445.

Gunn, A. S. and K. A. Tudhope. (2006) 'Ethical, Cultural and Spiritual Dimensions of Genetic Engineering'. Paper presented at the 5th Global Conference on Environmental Justice and Global Citizenship, Oxford, U.K., 3–6 July.

Gutteling, J. M. (2005) 'Mazur's Hypothesis on Technological Controversy and Media', *International Journal of Public Opinion Research*, 17, pp. 23–41.

Habermas, J. (1984) *The Theory of Communicative Action* (London: Heinemann).

Hajer, M. (1995) *The Politics of Environmental Discourse: Ecological Modernization and the Policy Process* (Oxford: Clarendon Press).

Hall, M. (1998) 'Ideas from Overseas: American Preservation and Italian Restoration', *The George Wright Forum*, 15(2), pp. 24–29.

———(2001). 'Restored Places or Restored Pasts: Managing Parklands in the Adirondacks and the Apennines'. Paper presented at the 1st International Conference of the European Society for Environmental History (ESEH), St Andrews, Scotland, 5–9 September.

Hallman, W. K. (2000) *Consumer Concerns About Biotechnology: International Perspectives*. Publication No. RR-0602-003. New Brunswick, NJ: Food Policy Institute.

Hallman, W. K. and W. C. Hebden. (2005) 'American Opinions of GM Food: Awareness, Knowledge, and Implications for Education', *Choices*, 20(4), pp. 239–242.

Hallman, W. K., C. L. Cuite and X. K. Morin. (2013) *Public Perceptions of Labeling Genetically Modified Foods*. Working Paper 2013–01. New Brunswick, NJ: Rutgers University. Available at humeco.rutgers.edu/documents_PDF/news/GMlabelingperceptions.pdf (accessed 11/01/2014).

Hallman, W. K., W. C. Hebden, H. L. Aquino, C. L. Cuite and J. T. Lang. (2003) *Public Perceptions of Genetically Modified Foods: A National Study of American Knowledge and Opinion*. Publication No. RR-1003-004. New Brunswick, New Jersey: Food Policy Institute.

Hallman, W. K., W. C. Hebden, C. L. Cuite, H. L. Aquino and J. T. Lang. (2004) *Americans and GM Food: Knowledge, Opinion and Interest in 2004*. Publication No. RR-1104-007. New Brunswick, New Jersey: Food Policy Institute.

Hampel, J., P. Grabner, H. Torgersen, D. Boy, A. Allansdottir, E. Jelsøe and G. Sakellaris. (2006) 'Public Mobilization and Policy Consequences', in G. Gaskell and M. W. Bauer (eds.) *Genomics and Society: Legal, Ethical and Social Dimensions* (London: Earthscan), pp. 77–94.

Haniotis, T. (2001) 'The Economics of Agricultural Biotechnology: Differences and Similarities in the US and the EU', in G. Nelson (ed.) *Genetically Modified Organisms in Agriculture: Economics and Politics* (San Diego: Academic Press), pp. 171–179.

Harris, M. (1968) *The Rise of Anthropological Theory: A History of Theories of Culture* (London: Routledge & K.P.).

Harvey, M., S. Quilley and H. Beynon. (2002) *Exploring the Tomato: Transformations of Nature, Society and Economy* (Cheltenham: Edward Elgar).

Hayes, G. (2007) 'Collective Action and Civil Disobedience: The Anti-GMO Campaign of the Faucheurs Volontaires', *French Politics*, 5(4), pp. 293–314.

Heiman, A. and D. Zilberman. (2011) 'The Effects of Framing on Consumers' Choice of GM Foods', *AgBioForum*, 14(3), pp. 171-179.

Heller, A. (1986) *Everyday Life* (London: Routledge).

Heller, C. (2004) 'Risky Science and Savoir-Faire: Peasant Expertise in the French Debate over Genetically Modified Crops', in M. E. Lien and B. Nerlich (eds.) *The Politics of Food* (Oxford: Berg), pp. 81–99.

——— (2006) 'Post-industrial 'Quality Agricultural Discourse': Techniques of Governance and Resistance in the French Debate over GM Crops', *Social Anthropology*, 14(3), pp. 319–334.

——— (2007) 'Techne Versus Technoscience: Divergent (and Ambiguous) Notions of Food "Quality" in the French Debate over GM Crops', *American Anthropologist*, 109(4), pp. 603–615.

Herlitz, L. (1997) 'Art and Nature in Pre-Classical Economics of the Seventeenth and Eighteenth Centuries', in M. Teich, R. Porter and B. Gustafsson (eds.) *Nature and Society in Historical Context* (Cambridge: Cambridge University Press), pp. 163–175.

Herrick, C. B. (2005) '"Cultures of GM": Discourses of Risk and Labelling of GMOs in the UK and the EU', *Area*, 37(3), pp. 286–294.

Hill, R. and C. Sendashonga. (2006) 'Conservation Biology, Genetically Modified Organisms, and the Biosafety Protocol', *Conservation Biology*, 20(6), pp. 1620–1625.

Hoban, T. J. (1997) 'Consumer Acceptance of Biotechnology: An International Perspective', *Nature Biotechnology*, 15, pp. 232–234.

——— (2004) *Public Attitudes Towards Agricultural Biotechnology*. ESA Working Paper No. 04–09. The Food and Agriculture Organization of the United Nations (FAO).

Hoban, T. J. and P. A. Kendall. (1992) *Consumer Attitudes About the Use of Biotechnology in Agriculture and Food Production*. Report for the U.S. Department of Agriculture – Extension Service.

Hood, C., H. Rothstein and R. Baldwin. (2001) *The Government of Risk: Understanding Risk Regulation Regimes* (Oxford: Oxford University Press).

Hopkinson, J. and H. Bottemiller Evich. (2014) 'Food Industry to Fire Preemptive GMO Strike'. *Politico.com*, 7 January. Available at http://www.politico.com/story/2014/01/gmo-labeling-bill-101853.html (accessed 12/05/2014).

House, J. (2000) 'The French Nineteenth-Century Landscape', in K. Flint and H. Morphy (eds.) *Culture, Landscape, and the Environment* (Oxford: Oxford University Press), pp. 131–148.

Hoyau, P. (1980) 'L'année du Patrimoine ou la Société de Conservation', *Les Révoltes Logiques*, 12, pp. 70–77.

Hubbard, K. and N. Hassanein. (2013) 'Confronting Coexistence in the United States: Organic Agriculture, Genetic Engineering, and the Case of Roundup Ready®Alfalfa', *Agriculture and Human Values*, 30(3), pp. 325–335.

Hughes, J. D. (2005) *The Mediterranean: An Environmental History* (Santa Barbara, CA: ABC-Clio).

Huntington, S. (1996) *The Clash of Civilizations and the Remaking of World Order* (New York: Simon & Schuster).

International Food Information Council (IFIC). (2006). *Food Biotechnology: A Study of U.S. Consumer Attitudinal Trends*. Available at http://www.foodinsight.org/Resources/Detail.aspx?topic=Food_Biotechnology_A_Study_of_U_S_Consumer_Attitudinal_Trends_2006_REPORT_ (accessed 09/05/2014).

—— (2010) *2010 'Consumer Perceptions of Food Biotechnology' Survey*. Available at http://www.foodinsight.org/Content/3843/Final_Executive%20Summary

%20Food%20Tech%20Report_Website%20version_7-7-10.pdf (accessed 04/05/2014).

——— (2012) *IFIC Consumer Perceptions of Food Technology Survey*. Available at http://www.foodinsight.org/Content/5519/IFIC%202012%20Food%20Technology%20Survey-US%20Topline%20Summary.pdf (accessed 04/05/2014).

Imort, M. (2005) 'A Sylvan People: Wilhelmine Forestry and the Forest as a Symbol of Germandom', in T. Lekan and T. Zeller (eds.) *Germany's Nature: Cultural Landscapes and Environmental History* (New Brunswick, NJ: Rutgers University Press), pp. 55–80.

Inghelbrecht, L., J. Dessein and G. Van Huylenbroeck. (2014) 'The Non-GM Crop Regime in the EU: How Do Industries Deal with This Wicked Problem?' *NJAS – Wageningen Journal of Life Sciences*, in press.

Jackson, P. T. (1999) 'Civilization on Trial', *Millennium: Journal of International Studies*, 28(1), pp. 141–153.

Jacobsen, E. (2004) 'The Rhetoric of Food: Food as Nature, Commodity and Culture', in M. E. Lien and B. Nerlich (eds.) *The Politics of Food* (Oxford: Berg), pp. 59–78.

Jacobsen, J. K. (2003) 'Duelling Constructivisms: A Post-mortem on the Ideas Debate in Mainstream IR/IPE', *Review of International Studies*, 29, pp. 39–60.

James, C. (2014) 'Global Status of Commercialized Biotech/GM Crops: 2013'. International Service for the Acquisition of Agri-Biotech Applications (ISAAA). Available at http://www.isaaa.org/resources/publications/briefs/46/executivesummary/default.asp (accessed 17/05/2014).

Jasanoff, S. (1995) 'Product, Process or Programme: Three Cultures and the Regulation of Biotechnology', in M. W. Bauer (ed.) *Resistance to New Technology: Nuclear Power, Information Technology and Biotechnology* (Cambridge: Cambridge University Press), pp. 311–331.

——— (2000) 'Commentary: Between Risk and Precaution', *Journal of Risk Research*, 3(3), pp. 277–282.

——— (2005) *Designs on Nature: Science and Democracy in Europe and the United States* (Princeton: Princeton University Press).

——— (2006) 'Biotechnology and Empire: The Global Power of Seeds and Science', *OSIRIS*, 21, pp. 273–292.

Jefferson, V. (2006) 'The Ethical Dilemma of Genetically Modified Food', *Journal of Environmental Health*, 69(1), pp. 33–34.

Jerome, N. W. (1977) 'Taste Experience and the Development of a Dietary Preference for Sweet in Humans: Ethnic and Cultural Variations in Early Taste Experience', in J. M. Weiffenbach (ed.) *Taste and Development* (Washington, DC: U.S. Government), pp. 235–248.

Johansen, T. R. (2007) 'Guest Editorial: Multifunctional Agriculture – Real Concerns, or Just an Excuse', *EuroChoices*, 6(3), pp. 6–12.

Kemp, D. D. (2004) *Exploring Environmental Issues: An Integrated Approach* (London: Routledge).

Kirkham, G. (2006) ' "Playing God" and "Vexing Nature": A Cultural Perspective', *Environmental Values*, 15(2), pp. 173–195.

Kiss, I. N. (1990) 'Sylviculture and Forest Administration in Hungary', in P. Brimblecombe and C. Pfister (eds.) *The Silent Countdown: Essays in European Environmental History* (Berlin: Springer-Verlag), pp. 106–124.

Klintman, M. and A. Kronsell. (2010) 'Challenges to Legitimacy in Food Safety Governance? The Case of the European Food Safety Authority (EFSA)', *European Integration*, 32(3), pp. 309–327.

Kniazeva, M. (2005) 'Marketing "Frankenfood": Appealing to Hearts or Minds?' *Journal of Food Products Marketing*, 11(4), pp. 21–39.

Knight, A. J. (2009) 'Perceptions, Knowledge and Ethical Concerns with GM Foods and the GM Process', *Public Understanding of Science*, 18(2), pp. 177–188.

Knowles, L. (2001). 'Bridging the (Agri)Cultural Divide: Public Perceptions of Agricultural Biotechnology in Europe and North America'. Paper presented at the Third Congress of the European Society for Agricultural and Food Ethics (EurSafe), Florence, Italy, 3–5 October.

Knowlton, L. (2007) 'Reading American Fat in France: Obesity and Food Culture', *European Journal of American Studies*, 2(2). Available at http://ejas.revues.org/document1363.html (accessed12/08/2014).

Kopicki, A. (2013) 'Strong Support for Labeling Modified Foods', *New York Times*, 27 July. Available at http://www.nytimes.com/2013/07/28/science/strong-support-for-labeling-modified-foods.html (accessed 12/08/2014).

Kratochwil, F. (1996) 'Is the Ship of Culture at Sea or Returning?' in Y. Lapid and F. Kratochwil (eds.) *The Return of Culture and Identity in IR Theory* (Boulder, CO: Lynne Rienner), pp. 201–222.

Krenzler, H. G. and A. MacGregor. (2000) 'GM Food: The Next Major Transatlantic Trade War?' *European Foreign Affairs Review*, 5, pp. 287–316.

Kunstler, J. H. (1998) *Home from Nowhere: Remaking Our Everyday World for the Twenty-First Century* (New York: Simon & Schuster).

Kurzer, P. and A. Cooper. (2007) 'What's for Dinner? European Farming and Food Traditions Confront American Biotechnology', *Comparative Political Studies*, 40(9), pp. 1035–1058.

Lang, J. T. (2013) 'Elements of Public Trust in the American Food System: Experts, Organizations, and Genetically Modified Food', *Food Policy*, 41, pp. 145–154.

Lang, J. T. and W. K. Hallman. (2005) 'Who Does the Public Trust? The Case of Genetically Modified Food in the United States', *Risk Analysis*, 25(5), pp. 1241–1252.

Lang, T. and M. Heasman. (2004) *Food Wars: The Global Battle for Mouths, Minds and Markets* (London: Earthscan).

Lassen, J. and A. Jamison. (2001) 'NGOs as Conveyors of Ethics in the Policy Process'. Paper presented at the Third Congress of the European Society for Agriculture and Food Ethics (EurSafe), Florence, Italy, 3–5 October.

Lassen, J., K. H. Madsen and P. Sandøe. (2002) 'Ethics and Genetic Engineering – Lessons to be Learned from GM foods', *Bioprocess and Biosystems Engineering*, 24, pp. 263–271.

Lawson, S. (2006) *Culture and Context in World Politics* (Houndmills: Palgrave).

Leitch, A. (2003) 'Slow Food and the Politics of Pork Fat: Italian Food and European Identity', *Ethnos*, 68(4), pp. 437–462.

Lekan, T. (2003) 'Seeking the Middle Ground: Landscape Preservation and Local Knowledge of Nature in the Rhine Valley, 1880–1939'. Paper presented at the 2nd International Conference of the European Society for Environmental History (ESEH), Prague, Czech Republic, 3–7 September.

Lekan, T. and T. Zeller. (2005) 'Introduction: The Landscape of German Environmental History', in T. Lekan and T. Zeller (eds.) *Germany's Nature: Cultural*

Landscapes and Environmental History (New Brunswick, NJ: Rutgers University Press), pp. 1–14.
Leopold, A. (1949) *A Sand County Almanac* (London: Oxford University Press).
Levenstein, H. A. (1988) *Revolution at the Table: The Transformation of the American Diet* (New York: Oxford University Press).
Levidow, L. (2000) 'Pollution Metaphors in the UK Biotechnology Controversy', *Science as Culture*, 9(3), pp. 325–351.
Levidow, L. and K. Boschert. (2008) 'Coexistence or Contradiction? GM Crops Versus Alternative Agricultures in Europe', *Geoforum*, 39, pp. 174–190.
Levidow, L. and J. Murphy. (2003) 'Reframing Regulatory Science: Trans-Atlantic Conflicts over GM Crops', *Cahiers d'économie et sociologie rurales*, 68–69, pp. 47–74.
Levidow, L. and S. Carr. (1997) 'How Biotechnology Regulation Sets a Risk/Ethics Boundary', *Agriculture and Human Values*, 14, pp. 29–43.
Levy, D. L. and P. Newell. (2000) 'Oceans Apart? Business Responses to Global Environmental Issues in Europe and the United States', *Environment*, 42(9), pp. 8–20.
Lieberman, S. and T. Gray. (2006) 'The So-called "Moratorium" on the Licensing of New Genetically Modified (GM) Products by the European Union 1998–2004: A Study in Ambiguity', *Environmental Politics*, 15(4), pp. 592–609.
Lien, M. E. (2004) 'The Politics of Food: An Introduction', in M. E. Lien and B. Nerlich (eds.) *The Politics of Food* (Oxford: Berg), pp. 1–17.
Listerman, T. (2010) 'Framing of Science Issues in Opinion-leading News: International Comparison of Biotechnology Issue Coverage', *Public Understanding of Science*, 19(1), pp. 5–15.
Lowenthal, D. (1991) 'British National Identity and the English Landscape', *Rural History*, 2(2), pp. 205–230.
——— (2000) 'Environment as Heritage', in K. Flint and H. Morphy (eds.) *Culture, Landscape, and the Environment* (Oxford: Oxford University Press), pp. 197–217.
Lusk, J. L. and A. Rozan. (2005) 'Consumer Acceptance of Biotechnology and the Role of Second Generation Technologies in the USA and Europe', *Trends in Biotechnology*, 23(8), pp. 386–387.
Lusk, J. L., W. B. Traill, L. O. House, C. Valli, S. R. Jaeger, M. Moore and B. Morrow (2006) 'Comparative Advantage in Demand: Experimental Evidence of Preferences for Genetically Modified Food in the United States and European Union', *Journal of Agricultural Economics*, 57(1), pp. 1–21.
Lynch, D. and D. Vogel. (2001) 'The Regulation of GMOs in Europe and the United States: A Case-Study of Contemporary European Regulatory Politics'. Council on Foreign Relations. Available at http://www.cfr.org/agricultural-policy/regulation-gmos-europe-united-states-case-study-contemporary-european-regulatory-politics/p8688 (accessed 09/05/2014).
Macilwain, C. (2003) 'Against the Grain', *Nature*, 422(6928), pp. 111–112.
Macnaghten, P. (2004) 'Animals in Their Nature: A Case Study of Public Attitudes to Animals, Genetic Modification and "Nature"', *Sociology*, 28(3), pp. 533–551.
Macnaghten, P. and J. Urry. (1998) *Contested Natures* (London: Sage).
Manning, F. C. R. (2000) 'Biotechnology: A Scientific Perspective', in A. Russell and J. Vogler (eds.) *The International Politics of Biotechnology: Investigating Global Futures* (Manchester: Manchester University Press), pp. 13–29.

March, J. G. and J. P. Olsen. (1984) 'The New Institutionalism: Organizational Factors in Political Life', *The American Political Science Review*, 78(3), pp. 734–749.

Marris, C., B. Wynne, P. Simmons and S. Weldon. (2001) *Public Perceptions of Agricultural Biotechnologies in Europe*. Final Report of the PABE Research Project.

McCauley, D. (2011) 'Bottom-Up Europeanization Exposed: Social Movement Theory and Non-state Actors in France', *Journal of Common Market Studies*, 49(5), pp. 1019–1042.

McCluskey, J. and J. Swinnen. (2011) 'The Media and Food-risk Perceptions', *EMBO Reports*, 12(7), pp. 624–629.

McKibben, B. (2003) *Enough: Genetic Engineering and the End of Human Nature* (London: Bloomsbury).

Melvin, J. (2009) 'USDA Unable to Weed out Unapproved Modified Foods'. *Reuters*. Available at http://www.reuters.com/article/2009/01/14/us-usda-gmo-policies-idUSTRE50D5ZD20090114 (accessed 04/05/2014).

Merchant, C. (1989) *Ecological Revolutions: Nature, Gender, and Science in New England* (Chapel Hill: University of North Carolina Press).

Mielby, H., P. Sandøe and J. Lassen. (2013) 'The Role of Scientific Knowledge in Shaping Public Attitudes to GM Technologies', *Public Understanding of Science*, 22(2), pp. 155–168.

Miller, H. I. (1998) 'The Emotional Response to Risks: Inevitable but Not Unmanageable', *AgBioForum*, 1(1), pp. 14–16.

Miller, S. (1995) 'Urban Dreams and Rural Reality: Land and Landscape in English Culture, 1920–45', *Rural History*, 6(1), pp. 89–102.

Mini, S. (2005) 'Genetics and Biotechnologies in Italian Mass Media', *Journal of Science Communication*, 4(3), pp. 1–13.

Mishler, W. and D. Pollack. (2003) 'On Culture, Thick and Thin: Toward a Neo-Cultural Synthesis', in D. Pollack and J. Jacobs (eds.) *Political Culture in Post-Communist Europe* (Farnham: Ashgate), pp. 237–257.

Monteiro, T. L. M. (2005) 'Preserving Europe's Heritage: Biodiversity, Landscape, and Agri-Cultural Policy in a Confederated Europe', *Environmental Law Reporter: News & Analysis*, 35(1), pp. 10065–10077.

Montpetit, E. and C. Rouillard. (2008) 'Culture and the Democratization of Risk Management: The Widening Biotechnology Gap Between France and Canada', *Administration & Society*, 39(8), pp. 907–930.

Montpetit, E., F. Varone and C. Rothmayr. (2007) 'Regulating ART and GMOs in Europe and North America: A Qualitative Comparative Analysis', in E. Montpetit, C. Rothmayr and F. Varone (eds.) *The Politics of Biotechnology in North America and Europe* (Lanham: Lexington Books), pp. 263–286.

Morgan, K., T. Marsden and J. Murdoch. (2006) *Worlds of Food: Place, Power and Provenance in the Food Chain* (Oxford: Oxford University Press).

Murdoch, J. and M. Miele. (1999) '"Back to Nature": Changing "Worlds of Production" in the Food Sector', *Sociologica Ruralis*, 39(4), pp. 465–483.

Murphy, J. and L. Levidow. (2006) *Governing the Transatlantic Conflict over Agricultural Biotechnology* (London: Routledge).

Myllyntaus, T. (2003) 'Writing the Past with Green Ink: The Emergence of Finnish Environmental History'. Available at https://www.academia.edu/984893/Writing_about_the_Past_with_Green_Ink_The_Emergence_of_Finnish_Environmental_History (accessed 11/05/2014).

Nash, R. (1967) *Wilderness and the American Mind* (New Haven: Yale University Press).
——— (1993) 'The Value of Wilderness', in C. Merchant (ed.) *Major Problems in American Environmental History* (Lexington, MA: D.C. Heath), pp. 395–409.
Nature. (2011) 'Editorial – Growing pains', *Nature*, 475(7356), pp. 265–266.
Nature News. (2013) 'US Regulation Misses Some GM Crops'. 20 August. Available at http://www.nature.com/news/us-regulation-misses-some-gm-crops-1.13580 (accessed 05/05/2014).
Navdanya International. (2011) *The GMO Emperor Has No Clothes*. A Global Citizens Report on the State of GMOs. Available at http://www.navdanya.org/attachments/Latest_Publications9.pdf (accessed 17/05/2014).
Nerlich, B. (2004) 'Risk, Blame and Culture: Foot and Mouth Disease in the Debate About Cheap Food', in M. E. Lien and B. Nerlich (eds.) *The Politics of Food* (Oxford: Berg, pp. 39–58.
Nisbet, M. C. and M. Huge. (2006) 'Attention Cycles and Frames in the Plant Biotechnology Debate: Managing Power and Participation through the Press/Policy Connection', *Harvard International Journal of Press/Politics*, 11(3), pp. 3–40.
Noelle-Neumann, E. (1993) *The Spiral of Silence: Public Opinion – Our Social Skin* (Chicago: University of Chicago Press).
Olson, M. (1971) *The Logic of Collective Action: Public Goods and the Theory of Groups* (Cambridge, MA: Harvard University Press).
Olwig, K. R. (1996) 'Reinventing Common Nature: Yosemite and Mount Rushmore – A Meandering Tale of Double Nature', in W. Cronon (ed.) *Uncommon Ground: Rethinking the Human Place in Nature* (New York: W.W. Norton), pp. 379–408.
O'Mahony, P. and T. Skillington. (1999) 'Constructing Difference: Discourse Coalitions on Biotechnology in the Press', in P. O'Mahony (ed.) *Nature, Risk and Responsibility: Discourses of Biotechnology* (Houndmills: Macmillan), pp. 100–113.
Pardo, R. and F. Calvo. (2006) 'Are Europeans Really Antagonistic to Biotech?', *Nature Biotechnology*, 24(4), pp. 393–395.
Patterson, L. A. (2000) 'Biotechnology Policy: Regulating Risks and Risking Regulation', in H. Wallace and W. Wallace (eds.) *Policy-Making in the European Union*, 4th edition (Oxford: Oxford University Press), pp. 317–344.
Payne, D. G. (1996) *Voices in the Wilderness: American Nature Writing and Environmental Politics* (Hanover, NH: University Press of New England).
Pelletier, D. L. (2006) 'FDA's Regulation of Genetically Engineered Foods: Scientific, Legal and Political Dimensions', *Food Policy*, 31, pp. 570–591.
Peters, H. P. and M. Sawicka. (2007) 'German Reactions to Genetic Engineering in Food Production', in D. Brossard, J. Shanahan and T. C. Nesbitt (eds.) *The Public, the Media and Agricultural Biotechnology* (Wallingford: CABI), pp. 57–96.
Peters, H. P., J. T. Lang, M. Sawicka and W. K. Hallman. (2007) 'Culture and Technological Innovation: Impact of Institutional Trust and Appreciation of Nature on Attitudes Towards Food Biotechnology in the USA and Germany', *International Journal of Public Opinion Research*, 19(2), pp. 191–220.
Pew Initiative on Food Biotechnology. (2006) *Memorandum from the Mellman Group, Inc.* Available at www.agronomica.org/Public%20Attitudes%20towards%20GM.doc (accessed 09/05/2014).

Phillips, A. (2001) 'The Nature of Cultural Landscapes: A Nature Conservation Perspective', in ICOMOS-UK (ed.) *The Cultural Landscape: Planning for a Sustainable Partnership Between People and Place* (London: International Council on Monuments and Sites UK), pp. 46–63.

Pilcher, J. (2006) *Food in World History* (New York: Routledge).

Pollack, M. A. (1996) 'The New Institutionalism and EC Governance: The Promise and Limits of Institutional Analysis', *Governance*, 9(4), pp. 429–458.

Pollack, M. A. and G. C. Shaffer. (2009) *When Cooperation Fails: The International Law and Politics of Genetically Modified Foods* (Oxford: Oxford University Press).

——— (2010) 'Biotechnology Policy: Between National Fears and Global Disciplines', in H. Wallace, M. A. Pollack and A. R. Young (eds.) *Policy-making in the European Union*, 6th edition (Oxford: Oxford University Press), pp. 331–356.

Poortinga, W. and N. F. Pidgeon. (2005) 'Trust in Risk Regulation: Cause or Consequence of the Acceptability of GM Food', *Risk Analysis*, 25(1), pp. 199–209.

Porter, R. (2000) ' "In England's Green and Pleasant Land": The English Enlightenment and the Environment', in K. Flint and H. Morphy (eds.) *Culture, Landscape, and the Environment* (Oxford: Oxford University Press), pp. 15–43.

Potter, D. M. (1954) *People of Plenty: Economic Abundance and the American Character* (Chicago: Chicago University Press).

Prakash, A. and K. L. Kollman. (2003) 'Biopolitics in the EU and the U.S.: A Race to the Bottom or Convergence to the Top?' *International Studies Quarterly*, 47, pp. 617–641.

Priest, S. H. (2000) 'US Public Opinion Divided over Biotechnology?' *Nature Biotechnology*, 18, pp. 939–942.

——— (2006) 'Public Discourse and Scientific Controversy: A Spiral-of-Silence Analysis of Biotechnology Opinion in the United States', *Science Communication*, 28(2), pp. 195–215.

Priest, S. H., H. Bonfadelli and M. Rusanen. (2003) 'The "Trust Gap" Hypothesis: Predicting Support for Biotechnology Across National Cultures as a Function of Trust in Actors', *Risk Analysis*, 23(4), pp. 751–766.

Puduri, V., R. Govindasamy, J. T. Lang and B. Onyango. (2005) 'I Will Not Eat It with a Fox; I Will Not Eat It in a Box: What Determines Acceptance of GM Food for American Consumers?' *Choices*, 20(4), pp. 257–261.

Radkau, J. (1997) 'The Wordy Worship of Nature and the Tacit Feeling for Nature in the History of German Forestry', in M. Teich, R. Porter and B. Gustafsson (eds.) *Nature and Society in Historical Context* (Cambridge: Cambridge University Press), pp. 228–239.

——— (2005) 'Germany as a Focus of European "Particularities" in Environmental History', in T. Lekan and T. Zeller (eds.) *Germany's Nature: Cultural Landscapes and Environmental History* (New Brunswick, NJ: Rutgers University Press), pp. 17–32.

Ramjoué, C. (2007) 'The Transatlantic Rift in Genetically Modified Food Policy', *Journal of Agricultural and Environmental Ethics*, 20(5), pp. 419–436.

Real Food Partners. (2013) *GMO-Free Finland Summit Explores the Benefits of Declaring Finland a GMO-Free Zone*. 20 August. Available at http://www.specialtyfood.com/media/uploads/article_files/NonGMOsummit_Stonyfield.pdf (accessed 13/05/2014).

Reeves, J. (2004) *Culture and International Relations: Narratives, Natives and Tourists* (London: Routledge).

Regal, P. J. (1999) 'A Brief History of Biotechnology Risk Debates and Policies in the United States'. Occasional Paper of the Edmonds Institute. Available at http://www.tc.umn.edu/~regal001/GEhistory.htm (accessed 04/05/2014).

Reisner, A. E. (2001) 'Social Movement Organizations' Reactions to Genetic Engineering in Agriculture', *American Behavioral Scientist*, 44(8), pp. 1389–1404.

Richardson, J. B. (2000) 'EU Agricultural Policies and Implications for Agrobiotechnology', *AgBioForum*, 3(2–3), pp. 77–83.

Rifkin, J. (1999) *The Biotech Century: How Genetic Commerce Will Change the World* (London: Phoenix).

——— (2006) 'This Crop Revolution May Succeed Where GM Failed', *The Guardian*, 26 October. Available at http://www.theguardian.com/commentisfree/2006/oct/26/food.comment (accessed 12/08/2014).

Roberts, B. K. (1998) 'Rural Settlement in Europe, 400–1500', in R. A. Butlin and R. A. Dodgshon (eds.) *An Historical Geography of Europe* (Oxford: Clarendon Press), pp. 54–72.

Roff, R. J. (2009) 'No Alternative? The Politics and History of Non-GMO Certification', *Agriculture and Human Values*, 26(4), pp. 351–363.

Rollins, W. (2000) *A Greener Vision of Home: Cultural Politics and Environmental Reform in the German Heimatschutz Movement, 1904–1918* (Ann Arbor: University of Michigan Press).

Rosendal, G. K. (2005) 'Governing GMOs in the EU: A Deviant Case of Environmental Policy-making?' *Global Environmental Politics*, 5(1), pp. 82–104.

Ross, M. H. (1997) 'Culture and Identity in Comparative Political Analysis', in M. I. Lichbach and A. S. Zuckerman (eds.) *Comparative Politics: Rationality, Culture, and Structure* (Cambridge: Cambridge University Press), pp. 42–80.

Rowe, F. (2001) 'The View from England', in ICOMOS-UK (ed.) *The Cultural Landscape: Planning for a Sustainable Partnership Between People and Place* (London: International Council on Monuments and Sites UK), pp. 122–130.

Rozin, P., C. Fischler and C. Shields-Argelès. (2012) 'European and American Perspectives on the Meaning of Natural', *Appetite*, 59(2), pp. 448–455.

Saba, A., A. Moles and L. J. Frewer. (1998) 'Public Concerns About General and Specific Applications of Genetic Engineering: A Comparative Study Between the UK and Italy', *Nutrition & Food Science*, 98(1), pp. 19–29.

Sagoff, M. (2004) 'Genetic Engineering and the Concept of the Natural', *Philosophy & Public Policy Quarterly*, 21(2/3), pp. 2–10.

Sanz-Lafuente, G. (2003). 'Economy and Nationalism: Forest Engineers and the Protection of the Woodland During the Franco Dictatorship'. Paper presented at the 2nd International Conference of the European Society for Environmental History (ESEH), Prague, Czech Republic, 3–7 September.

Sassatelli, R. and A. Scott. (2001) 'Novel Food, New Markets and Trust Regimes: Responses to the Erosion of Consumers' Confidence in Austria, Italy and the UK', *European Societies*, 3(2), pp. 213–244.

Sato, K. (2013) 'Genetically Modified Food in France: Symbolic Transformation and the Policy Paradigm Shift', *Theory and Society*, 42(5), pp. 477–507.

Sauer, C. O. (1963) *Land and Life: A Selection from the Writings of Carl Ortwin Sauer* (Berkeley: University of California Press).

Schaffer, S. (1997) 'The Earth's Fertility as a Social Fact in Early Modern Britain', in M. Teich, R. Porter and B. Gustafsson (eds.) *Nature and Society in Historical Context* (Cambridge: Cambridge University Press), pp. 124–147.

Scharfstein, B.-A. (1989) *The Dilemma of Context* (New York: New York University Press).

Scheese, D. (2002) *Nature Writing: The Pastoral Impulse in America* (New York: Routledge).

Scherzberg, A. (2006) 'EU-US Trade Disputes About Risk Regulation: The Case of Genetically Modified Organisms', *Cambridge Review of International Affairs*, 19(1), pp. 121–137.

Schilling, B. J., W. K. Hallman, A. O. Adelaja and L. J. Marxen. (2002) *Consumer Knowledge of Food Biotechnology: A Descriptive Study of U.S, Residents*. Publication No. RR-0502-002. New Brunswick, NJ: Food Policy Institute.

Schmidt, V. A. (2010) 'Taking Ideas and Discourse Seriously: Explaining Change Through Discursive Institutionalism as the Fourth "New Institutionalism"', *European Political Science Review*, 2(1), pp. 1–25.

Schmoll, F. (2005) 'Indication and Identification: On the History of Bird Protection in Germany, 1800–1918', in T. Lekan and T. Zeller (eds.) *Germany's Nature: Cultural Landscapes and Environmental History* (New Brunswick, NJ: Rutgers University Press), pp. 161–182.

Schultz, P. W. and L. Zelezny. (2003) 'Reframing Environmental Messages to be Congruent with American Values', *Human Ecology Review*, 10(2), pp. 126–136.

Schurman, R. and W. Munro. (2009) 'Targeting Capital: A Cultural Economy Approach to Understanding the Efficacy of Two Anti-Genetic Engineering Movements', *American Journal of Sociology*, 115(1), pp. 155–202.

Schurman, R. and W. A. Munro. (2010) *Fighting for the Future of Food: Activists Versus Agribusiness in the Struggle over Biotechnology* (Minneapolis: University of Minnesota Press).

Schweiger, T. G. (2001) 'Europe: Hostile Lands for GMOs', in B. Tokar (ed.) *Redesigning Life: The Worldwide Challenge to Genetic Engineering* (London: Zed Books), pp. 361–372.

Scott, I. M. (2000) 'Green Symbolism in the Genetic Modification Debate', *Journal of Agricultural and Environmental Ethics*, 13, pp. 293–311.

Seifert, F. (2004). 'The Transatlantic Conflict over Biotechnology and the Hegemony of Physical Risk'. Paper presented at the 45th Annual Isa Convention, Montreal, Quebec, Canada, Montreal, Quebec, Canada.

——— (2006) *Divided We Stand: The EU as Dissonant Player in the Global Governance of Agro-Food Biotechnology*. UNU-IAS Working Paper No. 146. Available at http://archive.ias.unu.edu/resource_centre/Seifert.pdf (accessed 12/05/2014).

——— (2009) 'Consensual NIMBYs, Contentious NIABYs: Explaining Contrasting Forms of Farmers GMO Opposition in Austria and France', *Sociologia Ruralis*, 49(1), pp. 20–40.

Shanahan, J., D. Scheufele and E. Lee. (2001) 'Attitudes About Agricultural Biotechnology and Genetically Modified Organisms', *Public Opinion Quarterly*, 65, pp. 267–281.

Sheingate, A. D. (2006) 'Promotion Versus Precaution: The Evolution of Biotechnology Policy in the United States', *British Journal of Political Science*, 36(2), pp. 243–268.

——— (2009) 'Federalism and the Regulation of Agricultural Biotechnology in the United States and European Union', *Journal of Comparative Policy Analysis*, 11(4), pp. 477–497.

Shepard, P. (2002) [1967] *Man in the Landscape: A Historic View of the Esthetics of Nature* (Athens, Georgia: University of Georgia Press).

Sierra Club (2001) 'Sierra Club Conservation Policies: Biotechnology'. Available at http://www.sierraclub.org/policy/conservation/biotech.asp (accessed 12/05/2014).

Sievert, J. (1998) 'Italy's Leap Forward in Nature Protection Legislation', *The George Wright Forum*, 15(2), pp. 43–49.

Simpson, J. (1995) *Spanish Agriculture: The Long Siesta, 1765–1965* (Cambridge: Cambridge University Press).

Sjöberg, L. (2000) 'Factors in Risk Perception', *Risk Analysis*, 20(1), pp. 1–11.

——— (2008) 'Genetically Modified Food in the Eyes of the Public and Experts', *Risk Management*, 10(3), pp. 168–193.

Skatssoon, J. (2005) 'Drought Prompts New Type of Mental Stress'. ABC Science Online, 24 May. Available at http://www.abc.net.au/science/news/stories/s1374921.htm (accessed 15/05/2014).

Skevas, T., E. M. Kikulwe, H. Papadopoulou, I. Skevas and J. Wesseler. (2012) 'Do European Union Farmers Reject Genetically Modified Maize? Farmer Preferences for Genetically Modified Maize in Greece', *AgBioForum*, 15(3), pp. 242–256.

Skillington, T. (1999) 'Modernity's Organic Economy of Governmentality', in P. O'Mahony (ed.) *Nature, Risk and Responsibility: Discourses of Biotechnology* (Houndmills: Macmillan), pp. 187–198.

Skogstad, G. (2006) 'Regulating Food Safety Risks in the European Union: A Comparative Perspective', in C. Ansell and D. Vogel (eds.) *What's the Beef? The Contested Governance of European Food Safety* (Cambridge, MA: MIT Press), pp. 213–236.

Skogstad, G. and E. Moore. (2004) 'Regulating Genetic Engineering in the United States and the European Union: Policy Development and Policy Resilience', *Policy and Society*, 24(4), pp. 32–56.

Sloan, T. S. (1999) 'The Colonization of the Lifeworld and the Destruction of Meaning', *Radical Psychology*, 1(2).

Slotkin, R. (1973) *Regeneration Through Violence: The Mythology of the American Frontier, 1600–1860* (Middletown, Conn.: Wesleyan University Press).

——— (1992) *Gunfighter Nation: The Myth of the Frontier in Twentieth-Century America* (New York: Atheneum).

Slovic, P. (1987) 'Perception of Risk', *Science*, 236(4799), pp. 280–285.

Smith, M. J. (2004) 'Mad Cows and Mad Money: Problems of Risk in the Making and Understanding of Policy', *British Journal of Politics & International Relations*, 6(3), pp. 312–332.

Smith, R. (2003) 'Prince Charles Is a GM Royal'. *The Mirror*, 16 October. Available at http://www.agbioworld.org/biotech-info/articles/biotech-art/princecharles.html#royal (accessed 12/08/2014).

Smythe, E. (2009) 'In Whose Interests? Transparency and Accountability in the Global Governance of Food: Agribusiness, the Codex Alimentarius, and the World Trade Organization', in J. Clapp and D. Fuchs (eds.) *Corporate Power in Global Agrifood Governance* (Cambridge, MA: MIT Press), pp. 93–123.

Solomon, R. C. (ed.) (2003) *What Is an Emotion? Classic and Contemporary Readings* 2nd edition (Oxford: Oxford University Press).

Stanziani, A. (2008) 'Defining "Natural Product" Between Public Health and Business, 17th to 21st Centuries', *Appetite*, 51, pp. 15–17.

Stephan, H. R. (2012) 'Revisiting the Transatlantic Divergence over GMOs: Toward a Cultural-political Analysis', *Global Environmental Politics*, 12(4), pp. 104–124.

Stirling, A. (2003) 'Risk, Uncertainty and Precaution: Some Instrumental Implications from the Social Sciences', in F. Berkhout, M. Leach and I. Scoones (eds.) *Negotiating Environmental Change: New Perspectives from Social Science* (Cheltenham: Edward Elgar), pp. 33–76.

Stokstad, E. (2006) 'A Kinder, Gentler Jeremy Rifkin Endorses Biotech, or Does He?' *Science*, 312(5780), pp. 1586–1587.

Stoll, M. (1997) *Protestantism, Capitalism, and Nature in America* (Albuquerque: University of New Mexico Press).

Swidler, A. (1986) 'Culture in Action: Symbols and Strategies', *American Sociological Review*, 51(2), pp. 273–286.

Swinnen, J. F. M. and T. Vandemoortele. (2010) 'Policy Gridlock or Future Change? The Political Economy Dynamics of EU Biotechnology Regulation', *AgBioForum*, 13(4), pp. 291–296.

Tarrow, S. (1998) *Power in Movement* (Cambridge: Cambridge University Press).

The Economist. (2013) 'Farming as Rocket Science'. 7 September. Available at http://www.economist.com/news/united-states/21584994-why-american-agriculture-different-european-variety-farming-rocket-science (accessed 17/05/2014).

Thomas, K. (1984) *Man and the Natural World: Changing Attitudes in England 1500–1800* (London: Penguin).

Thompson, P. B. (1997) 'Food Biotechnology's Challenge to Cultural Integrity and Individual Consent', *Hastings Center Report*, 27(4).

——— (2001) 'The Reshaping of Conventional Farming: A North American Perspective', *Journal of Agricultural and Environmental Ethics*, 14(2), pp. 217–229.

Thomson, J. and L. Dininni. (2005) 'What the Print Media Tell Us About Agricultural Biotechnology: Will We Remember?' *Choices*, 20(4), pp. 247–252.

Tiberghien, Y. (2009) 'Competitive Governance and the Quest for Legitimacy in the EU: the Battle over the Regulation of GMOs Since the mid-1990s', *Journal of European Integration*, 31(3), pp. 389–407.

Tilly, C. (1978) *From Mobilization to Revolution* (Reading, MA: Addison-Wesley).

Tokar, B. (ed.) (2001). *Redesigning Life? The Worldwide Challenge to Genetic Engineering* (London: Zed Books).

——— (2009) 'Toward Food Sovereignty in Vermont and Northern New England'. 19 May. Available at http://www.social-ecology.org/2009/05/toward-food-sovereignty-in-vermont-and-northern-new-england/ (accessed 15/05/2014).

Toke, D. (2004) *The Politics of GM Food: A Comparative Study of the UK, USA, and EU* (London: Routledge).

Toke, D. and D. Marsh. (2003) 'Policy Networks and the GM Crops Issue: Assessing the Utility of a Dialectical Model of Policy Networks', *Public Administration*, 81(2), pp. 229–251.

Torgersen, H., J. Hampel, M.-L. von Bergmann-Winberg, E. Bridgman, J. Durant, E. Einsiedel, B. Fjaestad, G. Gaskell, P. Grabner, P. Hieber, E. Jelsoe, J. Lassen, A. Marouda-Chatjoulis, T. H. Nielsen, T. Rusanen, G. Sakellaris, F. Seifert, C. Smink, T. Twardowski and M. W. Kamara (2002) 'Promise, Problems and

Proxies: Twenty-five Years of Debate and Regulation in Europe', in M. W. Bauer and G. Gaskell (eds.) *Biotechnology: The Making of a Global Controversy* (Cambridge: Cambridge University Press), pp. 21-94.

Touraine, A. (2004) 'Neo-Modern Ecology', *New Perspectives Quarterly*, 21(4), 129–133.

Trumbull, G. (2012) *Strength in Numbers: The Political Power of Weak Interests* (Cambridge, MA: Harvard University Press).

Tsioumani, E. (2004) 'Genetically Modified Organisms in the EU: Public Attitudes and Regulatory Developments', *RECIEL*, 13(3), pp. 279–288.

Tuan, Y.-F. (1974) *Topophilia: A Study of Environmental Perception, Attitudes, and Values* (Englewood Cliffs: Prentice-Hall).

USDA Foreign Agricultural Service. (2013) *EU-27 Agricultural Biotechnology Annual*. GAIN Report No. FR9142. 12 July.

van Dalen, W. (1997) 'European Debates in Bioethics: Diverse Topics and Procedures', *Biotechnology and Development Monitor*, 32, pp. 811.

van den Daele, W. (2007) 'Unease Finds Legitimate Expression in Risk'. Available at http://www.gmo-safety.eu/debate/495.unease-finds-legitimate-expression-risk.html (accessed 09/05/2014).

Van Waarden, F. (2006) 'Taste, Traditions, Transactions, and Trust: The Public and Private Regulation of Food', in C. Ansell and D. Vogel (eds.) *What's the Beef? The Contested Governance of European Food Safety* (Cambridge, MA: MIT Press), pp. 35–59.

Varone, F., C. Rothmayr and E. Montpetit. (2007) 'Comparing Biotechnology Policy in Europe and North America: A Theoretical Framework', in E. Montpetit, C. Rothmayr and F. Varone (eds.) *The Politics of Biotechnology in North America and Europe* (Lanham: Lexington Books), pp. 1–34.

Vicsek, L. (2013) ' "Gene-fouled or Gene-improved?" Media Framing of GM Crops and Food in Hungary', *New Genetics and Society*, 32(1), pp. 54–77.

Visser, M. (1999) 'Food and Culture: Interconnections', *Social Research*, 66(1), pp. 117–130.

Vogel, D. (2001) *The Protestant Ethic and the Spirit of Environmentalism: The Cultural Roots of Green Politics and Policies*. Unpublished Paper. Available at http://faculty.haas.berkeley.edu/vogel/paperprot.pdf (accessed 15/05/2014).

——— (2003) 'The Hare and the Tortoise Revisited: The New Politics of Consumer and Environmental Regulation in Europe', *British Journal of Political Science*, 33(4), pp. 557–580.

——— (2012) *The Politics of Precaution: Regulating Health, Safety, and Environmental Risks in Europe and the United States* (Princeton: Princeton University Press).

Vogler, J. and D. McGraw. (2000) 'An International Environmental Regime for Biotechnology', in J. Vogler and A. Russell. (eds.) *The International Politics of Biotechnology* (Manchester: Manchester University Press), pp. 123–141.

Vogt, D. U. (2001) *Food Biotechnology in the United States: Science, Regulation, and Issues*. Order Code RL30198. Congressional Research Service.

Wagner, W., N. Kronberger , G. Gaskell, N. C. Allum, A. Allansdottir, S. de Cheveigné, U. Dahinden, C. Diego, L. Montali, A. T. Mortensen, U. Pfenning, T. Rusanen and N. Seger (2001) 'Nature in Disorder: The Troubled Public of Biotechnology', in G. Gaskell and M. W. Bauer (eds.) *Biotechnology 1996–2000: The Years of Controversy* (London: Science Museum), pp. 80–95.

Wagner, W., N. Kronberger, N. Allum, S. de Cheveigné, C. Diego, G. Gaskell, M. Heinßen, C. Midden, M. Ødegaard, S. Öhman, B. Rizzo, T. Rusanen and A. Stathopoulou. (2002) 'Pandora's Genes: Images of Genes and Nature', in M. W. Bauer and G. Gaskell (eds.) *Biotechnology: The Making of a Global Controversy* (Cambridge: Cambridge University Press), pp. 244–278.

Wallwork, J. and J. A. Dixon. (2004) 'Foxes, Green Fields and Britishness: On the Rhetorical Construction of Place and National Identity', *British Journal of Social Psychology*, 43, pp. 21–39.

Walter, F. (1990) 'The Evolution of Environmental Sensitivity 1750–1950', in P. Brimblecombe and C. Pfister (eds.) *The Silent Countdown: Essays in European Environmental History* (Berlin: Springer-Verlag), pp. 231–247.

Weber, M. (1930) [1905] *The Protestant Ethic and the Spirit of Capitalism* (London: Unwin).

White Jr., L. (1972) 'The Expansion of Technology 500–1500', in C. M. Cipolla (ed.) *The Fontana Economic History of Europe: The Middle Ages* (London: Fontana), pp. 143–174.

Whited, T. L., J. I. Engels, R. C. Hoffman, H. Ibsen and W. Verstegen. (2005) *Northern Europe: An Environmental History* (Santa Barbara, CA: ABC-Clio).

Whyte, I. D. (1998) 'Rural Europe Since 1500: Areas of Retardation and Tradition', in R. A. Butlin and R. A. Dodgshon (eds.) *An Historical Geography of Europe* (Oxford: Clarendon Press), pp. 243–258.

Wiener, J. B., M. D. Rogers, J. K. Hammit and P. H. Sand. (eds.) (2011) *The Reality of Precaution: Comparing Risk Regulation in the United States and Europe* (Abingdon: Earthscan).

Wiener, M. J. (1981) *English Culture and the Decline of Industrial Spirit, 1850–1980* (Cambridge: Cambridge University Press).

Wilkins, L. (2001) 'A Primer on Risk: An Interdisciplinary Approach to Thinking About Public Understanding Of Agbiotech', *AgBioForum*, 4(3–4), pp. 163–172.

Williams, R. (1973) *The Country and the City* (London: Chatto & Windus).

——— (1976) *Keywords: A Vocabulary of Culture and Society* (London: Fontana Press).

Williams, R. H. (2004) 'The Cultural Context of Collective Action: Constraints, Opportunities, and the Symbolic Life of Social Movements', in D. A. Snow, S. A. Soule and H. Kriesi (eds.) *The Blackwell Companion to Social Movements* (Malden, MA: Blackwell), pp. 91–115.

Wilson, S. (1995) *Cultural Materialism: Theory and Practice* (Oxford: Blackwell).

Witoszek, N. (1997) 'The Anti-Romantic Romantics: Nature, Knowledge, and Identity in Nineteenth-Century Norway', in M. Teich, R. Porter and B. Gustafsson (eds.) *Nature and Society in Historical Context* (Cambridge: Cambridge University Press), pp. 209–227.

Wohlers, A. E. (2010) 'Regulating Genetically Modified Food: Policy Trajectories, Political Culture, and Risk Perceptions in the U.S., Canada, and EU', *Politics and the Life Sciences*, 29(2), pp. 17–39.

Worster, D. (1993) *The Wealth of Nature* (New York: Oxford University Press).

——— (1994a) *Nature's Economy: A History of Ecological Ideas* (Cambridge: Cambridge University Press).

——— (1994b) *An Unsettled Country: Changing Landscapes of the American West* (Albuquerque: University of New Mexico Press).

Wright, P. (1985) *On Living in an Old Country: The National Past in Contemporary Britain* (London: Verso).

Zajc, J. and K. Erjavec. (2012) ' "Othering" Agricultural Biotechnology: Slovenian Media Representation of Agricultural Biotechnology', *Public Understanding of Science*, OnlineFirst(in print).

Zechendorf, B. (1994) 'What the Public Thinks About Biotechnology', *Bio/Technology*, 12, pp. 870–875.

——— (1998) 'Agricultural Biotechnology: Why Do Europeans Have Difficulty Accepting It?' *AgBioForum*, 1(1), pp. 8–13.

Zurek, L. (2007) 'The European Communities Biotech Dispute: How the WTO Fails to Consider Cultural Factors in the Genetically Modified Food Debate', *Texas International Law Journal*, 42, pp. 345–368.

Zwart, H. (2003) 'Aquaphobia, Tulipmania, Biophilia: A Moral Geography of the Dutch Landscape', *Environmental Values*, 12, pp. 107–128.

Zwick, M. M. (1998) *Wertorientierungen und Technikeinstellungen im Prozeß gesellschaftlicher Modernisierung: Das Beispiel der Gentechnik*. Stuttgart: Akademie für Technikfolgenabschätzung in Baden-Würtemberg. Available at http://elib.uni-stuttgart.de/opus/volltexte/2004/1663/pdf/ab106.pdf (accessed 11/05/2014).

Index

Note: The letter 'n' following locators refers to notes.

CPSIA information can be obtained at www.ICGtesting.com
Printed in the USA
LVOW04*2309080515

437789LV00009B/238/P

9 780230 284579